DISCARD

EXPLORING CAREERS IN THE CONSTRUCTION INDUSTRY

EXPLORING CAREERS IN THE CONSTRUCTION INDUSTRY

By
ELIZABETH STEWART LYTLE

T 1004103

The Rosen Publishing Group, Inc.
NEW YORK

Published in 1992, 1995 by The Rosen Publishing Group, Inc.
29 East 21st Street, New York, NY 10010

Copyright 1992, 1995 by Elizabeth Stewart Lytle

All rights reserved. No part of this book may be reproduced in any form without permission in writing from the publisher, except by a reviewer.

Revised Edition 1995

Library of Congress Cataloguing-in-Publication Data
Lytle, Elizabeth Stewart.
  Careers in the Construction Industry/by Elizabeth Stewart Lytle.
    p. cm.
  Includes bibliographical references and index.
  Summary: Examines various jobs in the building trades, describing the training required, types of work involved, opportuinities for advancement, and more.
  ISBN 0-8239-1956-0
  1. Construction industry—Vocational guidance—Juvenile literature. 2. Building trades—Vocational guidance—Juvenile iterature. [Building trades—Vocational guidance.
  2. Vocational guidance.] I. Title.
HD9715.A2L97 1992
624'.023'73—dc20
                                                              91-43727
                                                                  CIP
                                                                   AC

Manufactured in the United States of America

For Chet, as enduring a sign of commitment and care as the homes you have built.

About the Author

Elizabeth Stewart Lytle is a photojournalist, teacher, and communications consultant in western Pennsylvania. She shares an interest in the construction trades and housing issues with her husband, a building contractor, and she is the author of three career books related to the construction trades.

As a free-lance magazine writer, she focuses on home design and improvement topics, along with travel articles. She has nearly 400 magazine article sales to her credit. As a contributing editor to *Your Home* and *Indoors & Out* magazines, her byline appears regularly in those pages. She has also written a number of career monographs and consumer articles for educational publishers in Chicago.

In pursuit of travel adventure stories, she has tried her hand at flying a soaring plane and has twice sailed the Saguenay River Fjord region of Quebec, Canada, to observe and photograph the summer migration of whales.

Mrs. Lytle has received several writing awards as a newspaper journalist and as a free-lance magazine writer. She also served three terms as a state director of Pennsylvania Women's Press Association.

She has done considerable work in the Writer's Workshop at the University of Iowa, where she earned a degree in liberal studies and is a member of Alpha Sigma Lambda national honorary society. She is currently working on a master's degree in communications.

About the Author

In 1991 the Franklin Area School District nominated Mrs. Lytle for the Sallie Mae First-Year Teacher Award, a program recognizing 100 of the nation's most outstanding new educators.

She and her husband are parents of one son, Christian.

Acknowledgments

Completion of this book would not have been possible without the interest and support of many highly skilled persons who gave generously of their time and talents. Particular thanks go to: J. Roger Glunt, incoming president of the National Association of Home Builders, the NAHB staff, and the Home Builders Institute, especially for research and photo support; Andrew L. Westley, Technical Assistance Coordinator for the Job Corps Program of the International Brotherhood of Painters & Allied Trades, for photo support; the Construction Labor Research Council of Washington, D.C.; the National Association of Women in Construction; the Job Corps; the Joint Apprenticeship Committees of the International Brotherhood of Electrical Workers and the International Brotherhood of Carpenters and Joiners of America; the U.S. Department of Labor, Employment and Training Administration; and Loren Adams, communications director of Marvin Windows, Minneapolis.

Contents

Foreword	viii
1. Hallmarks of the Construction Industry	1
2. The Specific Trades	16
3. Union or Independent Craftsworker?	59
4. Opportunities for Women and Minorities	63
5. Education and Training	68
6. Entering the World of Work	86
7. Opportunities for Advancement	93
8. Details to Consider	99
9. Becoming Your Own Boss	106
10. A Day in the Life of . . .	116
11. Toward the Future	144
Glossary	154
Appendix	158
For Further Reading	172
Index	175

Foreword

When I was younger and thinking about my future, there was never any doubt in my mind that I would become a homebuilder.

That was because I had seen firsthand by observing the construction industry (my dad owned a lumber supply business) how satisfying it is to build homes and help families achieve the dream of home ownership.

Not everyone has the opportunity to observe an industry closely as I did. But anyone—and that certainly includes women and minorities—who wants to work in the construction industry can find a niche in residential or commercial building.

There are numerous opportunities for young people with a basic high school education to receive specialized training in a construction trade. The need for skilled workers has never been greater than it is today. During the 1990s the construction industry will need to recruit more than 340,000 workers annually to fill new jobs and to replace people who leave the workforce.

Many colleges offer degrees in construction technology, and people with other degrees ranging from management to finance will find numerous career opportunities in the construction field.

Finally, the construction industry provides great opportunities for entrepreneurs with drive and determination who want to own their own business.

As a homebuilder for more than twenty-five years, I can honestly say that I have loved every minute of my career. I urge young women and men who are looking for a challenging and rewarding career to investigate opportunities in the construction field and consider making it their life's work.

—J. Roger Glunt
Vice President/Treasurer
National Association of Home Builders

1

Growth, Variety, and Accomplishment: Hallmarks of the Construction Industry

A dramatic city skyline with its high-rise office towers seems to have little in common with the pastoral quiet of a country town, yet both are composed of buildings created in essentially the same way. Regardless of locale, skilled workers erected the metal, lumber, and plastics that shelter families and provide places for commerce, learning, and the other activities of modern life.

A building is something that lasts. Across Europe, people continue to live and work in fine stone and timber structures built 900 years ago or more. On this side of the Atlantic, the oldest house in America still welcomes visitors to St. Augustine, Florida. The hands that raised those original walls began a long and proud tradition that you could join today, carrying the history of the construction trades into a new century.

Along with practical skills with which to earn a good living, the construction worker acquires a sense of quiet personal pride and accomplishment that stands as long as the building itself. Over the years, craftsworkers in the building trades have contributed to the architectural and economic history of this nation. They have also passed along the trade to a new generation. Through the tradition of apprenticeship training, thousands of young men and women are learning these trades, en route to secure, well-paid jobs. You could be one of them.

Right now more than four million men and women work in the various facets of America's building industry. This means that a third of the nation's total skilled labor force are represented in approximately a dozen construction job categories. That is more jobs than in any other national enterprise, including steelmaking and the automotive industry.

A Place for You

Growth, variety, and accomplishment—certainly these words describe the construction trades. In these pages you will find a comprehensive study of the building trades, America's construction industry, addressing the information needs of a young adult with career decisions to make. Your first question is likely to be, "Is there a good job for me?"

Yes, indeed, if you have the necessary skills and training, combined with a winning attitude. Perhaps the best news, if you don't particularly enjoy the classroom environment, is that in this industry there are still hundreds of rewarding careers that do not require a college degree.

If you enjoy working with your hands and mind, if you can follow oral or written directions, and if you can work as part of a team, the construction industry may be for you. Throughout the 1990s the construction industry must recruit more than 350,000 workers each year to fill new positions and replace workers who retire or change jobs. Start planning now to launch your own successful career.

In these pages you will learn about particular jobs and where to find them. You will also discover what it takes

HALLMARKS OF THE CONSTRUCTION INDUSTRY

to be hired in the construction industry, and how to keep your job. You will read about the benefits and rewards a worker can expect, along with opportunities for advancement. Finally, you will examine the sort of changes that experts think will affect the future of the constuction industry.

Along the way, you will discover what it is like to work in a big business with one thousand or more employees, or for your own one-person contracting company, since both are common in an industry marked by diversity and entrepreneurial spirit.

THE NATURE OF THE BUSINESS

When most of us think of construction, we think of new buildings and other facilities, yet the remodeling industry has grown to a billion-dollar part of the construction picture in North America. Another emerging source of employment for people skilled in construction trades is in apartment maintenance, particularly in urban and resort areas with their thousands of housing units in the form of condominiums, townhouses, and apartments. As the trend toward higher-density housing grows, thousands of jobs will require the expertise of men and women who can enlarge or modernize a building without jeopardizing structural elements already in place.

Construction workers do interesting work and are well paid for their efforts. The average union wage of all construction trades as of January 1994 stood at $26 per hour. Of course, pay scales vary from union to nonunion workers, from one geographic region of the country to another, and from one trade to the next. For several years now, construction plumbers have been the best-paid craftspeople in the industry. No matter what job category you mention, hourly wages in the building trades are among the highest paid to any skilled workers. In general, construction mechanics earn significantly more than manufacturing workers. During the peak building season when overtime hours are available, the pay scale for exceeding a 40-hour week is half-again or double the regular hourly rate. Even the training pay in construction trades exceeds training rates for other industries.

The National Association of Home Builders (NAHB) calls the need for craftsworkers in the building trades "tremendous," noting that the residential contractors of America have long held the lead in offering construction workers the most new jobs. The composition of tomorrow's workforce is changing, too. Through the 1990s women and minorities will make up more than a third of all newly hired construction workers. There has never been a better time for someone interested in a construction career to realize his or her dream.

Although critical labor shortages mean job openings, experts in the industry see an ominous cloud gathering on the horizon. Its name is workplace illiteracy, which means that too many of today's young workers are coming into the job market lacking basic educational skills. Even those who have a high school diploma cannot perform math functions well enough to do the job or cannot read directions or understand verbal instructions. Many of the new employees hired for construction jobs have skill levels below those expected for their jobs. As a group, recent high school graduates have been found to lack the ability to read, write, and reason at a level demonstrated by job applicants as recently as five years earlier.

The NAHB recently surveyed construction industry executives and found that two thirds of all firms surveyed are having difficulty recruiting skilled craft labor. Unfortunately, this shortfall in skills comes at a time when increased mechanization and new technology demand ever-higher levels of skill.

Hallmarks of the Construction Industry

The results range from disappointing to disastrous. Undereducated workers are frustrated by the demands of the work world and often have trouble keeping a job. Those who manage to stumble along make more mistakes, ruin more materials, and have more job-related accidents.

Remodeling and building and apartment maintenance, two of the fastest-growing segments of residential housing, both require workers with multiple skills, yet fewer and fewer qualified newcomers are available to fill these jobs.

The basic skills employers are concerned about go beyond the "three Rs." An industry report cited inadequacies "in thinking and problem-solving skills, as well as in self-discipline, reliability, and other personal traits that characterize dependable, responsible, and adaptable workers."

Citing workplace illiteracy as "one of today's most pressing problems," the Home Builders Institute, the educational arm of the NAHB, has been sponsoring locally based training programs since 1967. The NAHB also developed the Craft Math curriculum, teaching math concepts by using carpentry applications; and CommuniCraft, designed to improve verbal and reading skills with teaching materials based on real-life construction applications.

The lesson is clear: If you want to succeed, make the most of high school, vocational training, and other educational opportunities. People who can think clearly, solve problems, apply their skills, and demonstrate a winning attitude will have their choice of good jobs.

Training Opportunities

Every year thousands of first-rate job training opportunities are available in the construction trades, making it possible for young men and women to go straight

Hallmarks of the Construction Industry

from high school to journeyworker status without spending a penny on their education. Known as apprenticeship, these work-study programs for the blue-collar professions are considered among the best ways to prepare for a career.

Apprentices work alongside people who are the very best in the business, learning such high-paying jobs as construction plumber, electrician, carpenter, bricklayer, and operating engineer. Apprentices attend classes that often use the latest and most effective teaching methods in adult education.

Besides receiving three or four years of free training, apprentices are paid an hourly wage for their work, earning a raise every time they successfully complete six months of the program.

But what if you left high school without graduating? Are you automatically locked out of these high-paying jobs? Absolutely not. There are several alternative methods for gaining entry to the field, notably the federal Job Corps program, which lets people finish their education and prepare for a lifelong career at the same time. You'll find more about that subject in Chapter 5.

This book looks at vocational aptitude testing, vocational education, high school programs, formal and informal apprenticeships, trade school and community college programs, on-the-job training, and college courses for those who aspire to hold one of the top positions or to own and operate their own business within the construction industry.

You may also choose between working for a group of contractors and being a full-time employee of a

Though structurally sound, this brick apartment block required total rehabilitation, serving as a hands-on learning laboratory for apprentice construction workers (courtesy Home Builders Institute).

construction firm. An entire chapter is devoted to a blueprint for making that leap of faith from being an employee to running your own show. Starting your own business is not all "power and glory"—there are also pitfalls—but if you prepare yourself carefully and make wise use of resources you can go it alone with very little start-up cost compared to other industries. The phrase "go it alone" has particular significance in the home-building industry, where an amazing number of one-person companies operate out of the self-employed person's home.

You will also discover that many skilled construction workers alternate between working for a contractor and working as contractors themselves on smaller jobs. Consider, for example, that one of every three carpenters is self-employed at some time during his or her career.

A CRAFT WITH A PROUD HISTORY

The ability to build useful structures, roads, and facilities has been held in high regard since the dawn of history. Early civilizations began when people took crude stone axes in hand and built huts and enclosures to ward off harsh weather and their enemies. People skilled in the use of tools gained the respect of their peers, and this aspect of craftsmanship has continued down through the centuries. Stonecutters and skilled workers erected the classic wonders of the ancient Greek and Roman worlds, and construction workers were not far behind the conquering armies that spread Roman civilization throughout western Europe.

Construction workers played a vital role in America's early history. There were more woodwrights and house joiners among the passengers sailing to the New World aboard the *Mayflower* than any other skilled craftsmen. As settlers moved west, construction workers built the new towns, bridges, railroads, and ships. Industrialization and population growth built an industry that today includes such varied construction as military bases, suburban housing developments, urban business towers, sports arenas, airports, and the Alaska pipeline. Future construction workers may have jobs that are out of this world literally. Space stations that orbit the earth will require construction workers to build and maintain them. The location and materials may change, but the need for skilled hands will not.

Coping with Change

How will people like you, who are just choosing a career, cope with the change expected in the construction industry over the next twenty-five to fifty years? By cultivating your ability to adapt. And how does one do that? By developing a winning personal attitude. You can be a leader in your career field if you build a strong foundation of knowledge, show that you are willing to master new skills, and vow to do the best job possible with every task put before you.

TYPES OF JOBS

The building trades include work in the construction, maintenance, and remodeling of homes, apartments, factories, dams, bridges, highways, pipelines, and other structures. Skilled employees, called *journeyworkers*, are assisted in carrying out their work by apprentices, helpers, and laborers.

Journeyworkers are usually grouped in three basic job categories. Many occupational groups are involved in construction, but the major classifications are the following:

Structural, including bricklayers, carpenters, concrete masons, ironworkers, riggers, and stonemasons.

Mechanical, including electricians, elevator constructors, operating engineers, plumbers, pipefitters, and sheetmetal workers.

Finishing, including asbestos workers, floor-covering installers, lathers and plasterers, painters, roofers, and terrazo, tile, and marble installers.

All skilled craftspeople in the construction trades are important, and their work is essential. They bring to the job skill, training, and experience. Those destined to enjoy the greatest success are also adaptable and open to change, because change brings new tools, new building materials, and new technologies. Still, change does not eliminate the need for seasoned judgment and practiced skills. These are fulfilling careers to last a lifetime.

If the ring of a hammer or the whine of a saw blade brings you a chill of excitement, read on. If you like to step back for a moment and appreciate a job well done, you should find plenty of opportunity in the building trades. The satisfaction of creating something useful, perhaps even beautiful, that may endure for a century or more is all part of a day's work.

It may surprise you to know that only 53 percent of all construction trades workers are employed in the construction industry. People with construction skills are employed in such industries as manufacturing, transportation, mining, and agriculture. Only about half of those employed in construction are at work building homes.

As you plan for the future, what jobs are expected to be in greatest demand? The answer is given in an industry survey conducted by the NAHB:

Jobs in Greatest Demand, by U.S. Region
New England States: roofers.
Middle Atlantic States (New Jersey, New York, Pennsylvania): plumber, pipefitter, steamfitter helpers; drywall installers, heating, air-conditioning,

and refrigeration mechanics; roofers, and carpenters.
South Atlantic States (Delaware, District of Columbia, Florida, Georgia, Maryland, North Carolina, South Carolina, Virginia, West Virginia): brick masonry.
East South Central States (Alabama, Kentucky, Mississippi, Tennessee): roofers, heating, air-conditioning, and refrigeration mechanics, stonemasons.
West South Central States (Arkansas, Louisiana, Oklahoma, Texas): drywall finishers, electricians, powerline installer helpers.
East North Central States (Illinois, Indiana, Michigan, Ohio, Wisconsin): drywall installers, tapers, heating, air-conditioning, and refrigeration mechanics.
West North Central District (Iowa, Kansas, Minnesota, Missouri, Nebraska, North Dakota, South Dakota): roofers, heating, air-conditioning, and refrigeration mechanics, plumbers, pipefitters, steamfitters.
Mountain District (Arizona, Colorado, Idaho, Montana, Nevada, New Mexico, Utah, Wyoming): roofers' helpers.
Pacific District (Alaska, California, Hawaii, Oregon, Washington): roofers, roofers' helpers.

A Typical Work Day

Assuming you are the sort of person with the talent and drive to build a good educational foundation and develop a winning attitude, let's take a look at a construction worker's typical workday.

It starts early, before the sun is high and hot. Most construction sites are abuzz by 7 a.m. Workers in the South and Southwest may be on the job at dawn during the summer, wrapping up the day's work by 1:30 p.m. Others call it a day between 3 and 5 p.m. unless there is overtime work. The 40-hour week reigns, but expect to be offered evening and weekend hours during busy seasons.

In regions where winter snows interfere with building schedules, modern construction methods and temporary shelters have made many a work site a more comfortable place while extending the building season.

Still, a day's work in construction calls for strenuous and demanding physical labor. You rely on a combination of skills, experience, and problem-solving ability to meet the demands of your job. As for equipment, safety shoes, a hard hat, and safety goggles or a visor are standard issue. A craftsperson keeps his or her tools clean, sharp, and stowed in a toolbox when not in use. Workers take pride in their tools and rarely lend them. Tools are never treated carelessly or left lying around. You are expected to supply your own hand tools and perhaps some of the smaller power tools. The large, free-standing equipment is generally provided by the contractor.

If you have ever witnessed a great orchestra in performance, you may be reminded of that shared commitment to a goal while at the construction site, for the art of erecting a building with all its exterior and interior systems is similar to the work of performing a symphony. Carpenters, electricians, plumbers, masons, and machine operators are all working apace. Teamwork is important, for a schedule must be met. The basic elements of construction must be completed before the next phase of work can begin. Errors and delays upset the schedule and erode the contractor's profit.

Yes, there are coffee breaks and time for lunch, but they generally coincide with natural breaks in the rhythm of the work. Craftsworkers rarely leave the site during the workday. At lunch time they adjourn with their meal to a stack of lumber for a 30-minute break.

A certain hierarchy or pecking order prevails on the

Hallmarks of the Construction Industry

An apprentice is usually a young person who is supervised by someone older and more experienced, often a journeyworker or other supervisor (courtesy Home Builders Institute).

job site. The apprentice, usually a young person engaged in learning a trade, is the low person on construction's totem pole. Journeyworkers are by far the most numerous on a job site, followed by one or more supervisors, the job superintendent, and the contractor. Depending on the size of the project, there may be engineers, architects, and any number of subcontractors on the site.

Carpenters, electricians, and plumbers may be employed in a prefabrication shop or mill, where they usually work a 40-hour week. On a construction site, however, these and other craftsworkers can expect to work longer hours for short periods when necessary to meet deadlines.

Of course, there are the inevitable days with no work, caused by inclement weather, delays in receiving materials, or other complications peculiar to the industry. Union workers who are laid off from a job or have finished work on a project can report to the union hall for help in getting a new assignment. Many of these arrangements are made early on Monday mornings to avoid being put on a waiting list.

As you might imagine, contractors specifically request workers who have a reputation for good work. Unless the weather is a problem, a good construction worker can nearly always find work.

You can also see that union construction workers are more likely to have a longstanding relationship with the people at their union hall than with any one employer. However, some union workers spend years in the employ of a single contracting firm.

Suppose you are not a union member; how do you find work? Most review the newspaper classified ads, ask fellow construction workers for leads, contractors they have worked for in the past, or visit the local Job Service Office. Most nonunion work is in residential construction, maintenance work, and small commercial construction, where union representation is not strong.

Remember the problem-solving skills we said a construction worker needs? Many workers have adapted to the "hurry up and wait" aspects of seasonal work by literally saving for a rainy (or snowy) day. Unemployment compensation, augmented by money put aside during the building season, provides for living expenses during the off-season. Many construction workers use those weeks to catch up on family life or follow leisure pursuits. Some forgo unemployment benefits, spending their time contracting indoor work projects, or traveling to a Sunbelt state for short-term work assignments.

2

The Specific Trades

Regardless of the nature of a construction project, each building site is a hive of activity. It is not uncommon to see a dozen highly skilled craftspeople pursuing their work at the same time. In the midst of all this clamor and bustle you are likely to see bricklayers and cement masons wielding their trowels, carpenters at work with hammer and saw, electricians stringing wire or cable, and glaziers installing glass. Depending on the stage of construction, you might also see heating and air-conditioning installers at work, the insulating crew, painters, landscapers, or a half dozen others.

All these activities going on at once may not be music to your ears, yet the process of erecting and equipping a building actually does resemble the performance of a carefully rehearsed orchestra in many ways. Each craft plays a part in the overall scheme, with timing, quality workmanship, and proper materials all contributing to the success of the project.

This chapter takes a close-up look at the major career choices in the construction industry. We shall see what each worker actually does and discover the rewards, challenges, and hazards that accompany each of the major job categories in the construction industry, as well as forecasts of demand for such workers through the year 2000. The list is arranged alphabetically.

THE SPECIFIC TRADES

A word of explanation concerning the salary ranges given: Traditionally, construction wages in the U.S. have varied by region, with the industrialized Northeast and the Southwest Pacific regions leading. The Southeast and South Central regions have traditionally reported the lowest construction wages. It is important to understand that wage figures listed here represent a range or average for unionized workers and may not reflect what you would receive as an entry-level or independent craftsworker.

Bricklayers and Cement Masons

Although many modern buildings have a sophisticated, high-tech look, the basic tools and techniques of this trade have changed little over the centuries. The bricklayer's or cement mason's toolbox still contains such basic items as trowel, pointer, chisel, hammer, mason's line, rule, level, jointer, and drill.

Using cement or mortar, bricklayers and masons build floors, walls, walkways, and other structural elements from a wide range of rugged materials including natural stone, brick, block, tile, and manufactured products. Their work is seen on the inside and outside of buildings and may involve a solid or veneer finish. A bricklayer who can lay up a straight corner or follow a complex geometric pattern is considered skilled. Both crafts may involve repair or replacement activities, and fireplace construction is a related aspect.

Much of the work is done outdoors, requiring workers to stand, kneel, and bend for hours. Job opportunities exist across the nation, but each geographic area has its good and bad times in construction activity, with unemployment resulting when work is slow. About three of every ten bricklayers and stonemasons are self-employed. The others work for special trade, building, or general contractors.

Apprenticeship programs generally offer the most comprehensive training for this career, although many bricklayers and stonemasons learned their craft at the elbow of an experienced worker. Entry-level positions as helpers or apprentices include duties such as transporting materials, moving scaffolding, and mixing and carrying mortar.

A mason typically specializes in brick, block, stone, or cement work. Stonemasonry approaches the level of an art form, employing stones that vary in size and shape laid out in patterns or in a jigsaw-puzzle design. The stones are bonded together by applying mortar through a portable squeeze bag resembling the sort pastry chefs use to decorate cakes.

Other masons lay fire-hardened bricks of clay or blocks formed of concrete. The cement mason works primarily with concrete products poured into forms to create floors, walls, driveways, or street surfaces.

According to statistics published in 1990, the U.S. has approximately 160,000 masons on the job in a field that is growing in opportunities, since few people are training to take the jobs of those who retire. Although mechanization has reduced the need for unskilled laborers, workers with good manual dexterity who are skilled in the use of masonry tools and equal to the demands of a strenuous job will have a choice of positions.

Masons are precise and meticulous workers who waste no energy in getting the job done. They pay attention to detail and can follow plans, including intricate geometric layouts. At one time the length of a mason's workyear was determined by the whims of nature. Since concrete and mortar need a certain minimum temperature in which to harden, many construction jobs in colder climates shut down in the fall. Today, modern devices and updated construc-

The Specific Trades

tion techniques make possible affordable, temporary artificial environments in which the bricklayer and mason ply their trade nearly year-round.

Training programs for this career last from one to three years. It is still common for employers to hire unskilled workers as laborers or helpers working with experienced masons. Promising workers receive on-the-job training.

Apprenticeships are available during which you learn by actually using mason's tools. Classroom work focuses on the terminology of the trade and practice in manipulative skills. Apprentices learn to spread mortar and install joint heads during classroom building projects. As their skills progress, apprentices are usually assigned to perform repair work under a trained supervisor. Instruction in reading blueprints, performing layout work, setting door and window frames, and cleaning masonry are also part of the apprenticeship package.

The satisfaction of being a bricklayer or mason is both quick and enduring. At the end of each day's work these craftspeople see the product of their labors. They also know that fifty years in the future they can bring their grandchildren to see that day's work.

Between 1986 and 1993 bricklayers' and cement masons' wages ranged from $15.70 to $32.09 per hour, according to the Construction Labor Research Council.

Carpenters

Carpenters belong to one of the oldest and most respected trades in the world. If you like working with tools and enjoy creating things, this field may be for you. In addition to construction carpenters, there are cabinetmakers, millwrights, and a relatively new specialist, the interior systems carpenter.

Carpenters play an important role in practically every type of construction activity. They are of two basic kinds, carpenters who perform "rough" work and those who do the "finish" work. A skilled carpenter can do both, often staying on a particular job until its completion.

Rough carpentry involves erecting the wood framework for buildings, including the subfloors, partitions, and floor joists. Rough carpenters also erect scaffolds and temporary structures needed at construction sites and build the forms that enclose newly poured concrete while it hardens. The work of the rough-in carpenter comes first in any construction project; important as it is to structural integrity, however, little of this early work may be evident in the finished product, especially a building.

Finish carpenters install molding, wood paneling, cabinets, windows, and doors with their hardware. They also build stairways and lay floors. Sometimes the finish carpenter installs wallboard and such flooring materials as resilient tile or sheet vinyl.

In small communities where highly specialized workers are less likely to be found, carpenters may hang windows and apply insulation, or finish the wood trim before they install it.

Carpenters work throughout the nation in communities of all sizes and in rural areas. The work is active and often physically demanding. Expect to stand, climb, kneel, or squat. It requires care and attention to detail to avoid falls and injury from sharp or rough materials. Respect for sharp tools and power equipment is also a daily demand.

These workers must read and interpret architectural blueprints or carefully follow instructions given by their supervisors, to insure that the final product is built according to plan. Materials and techniques used must conform to local building codes.

Carpenters use both hand and power tools, and safety

The Specific Trades

is of utmost concern. Most carpenters demonstrate a real affinity for the tools with which they earn a living. Most have their favorite hand tools including hammers, saws, chisels, and planes, each kept in tip-top condition, inscribed with the owner's identifying mark, and taken home at the end of the day. Commonly used power tools include portable saws, drills, power nailers, and impact tools, which may also belong to the carpenter, although the larger, more specialized power tools are sometimes supplied by the builder or contractor.

The majority of carpenters are employed by contractors and builders who construct new buildings or restore and renovate existing buildings. Although carpenters help to build office buildings, factories, warehouses, and shopping centers, an estimated 60 percent of all carpenters go into residential housing work. In areas where union membership is not strong, wages tend to be lower than in unionized areas. Home construction sites may also be marked by seasonal downturns in employment. For the most part, construction carpenters enjoy the rigorous yet satisfying nature of their work. Physical strength is required, and the work site is often outdoors. Travel is also possible.

Housing contractors are always seeking good residential carpenters. They are called the master builders because they are the primary craftsworkers on homes, apartments, condominiums, and the like. A significant number of carpenters are self-employed or combine working for wages with part-time business ventures of their own. Government agencies and manufacturing firms also hire carpenters, usually to perform repairs and maintenance.

As you might imagine, manual dexterity is vital, as is the ability to solve math problems quickly and accurately. A carpenter needs to be in good physical

The Specific Trades

condition, with a sense of balance and no fear of heights.

In the past many people acquired skills by working as a carpenter's helper or for contractors who provided some training. Today the carpenter with a formal apprenticeship background is in great demand, earns the highest pay, and makes the most rapid advancement. Apprentices often start out carrying materials, keeping building sites clear, and learning about job safety and the use and care of power tools. More demanding tasks follow with demonstrated skills and ability. The industry generally agrees that a carpenter needs seven or more years of work experience to master this trade.

Many carpenters find work in the remodeling industry or in contract maintenance. The field of historic preservation, meaning historically accurate restoration of public and private buildings, is also growing as communities come to appreciate their architectural heritage. Most large industrial plants have either full-time construction craftspeople to perform repair and maintenance tasks or contractual agreements with outside firms to perform such work.

As of January 1993 union construction carpenters earned an average of $25.31 an hour. Annual earnings are not always a simple calculation of this high hourly rate, however. Weather, downtime between jobs, and other factors have a serious impact on wages.

One third of the carpenters in the U.S. are self-employed. Job opportunities over the next ten years should be abundant as population growth adds to demand for homes and other buildings. However, the housing industry remains extremely sensitive to the national economy.

Mastering the art of building stairs is a milestone in every carpenter's training (courtesy Home Builders Institute).

Carpenters advance more readily to general supervisory positions in the construction industry than do other craftspeople because they are involved in the entire construction process. For the same reason, carpenters are more likely than other craftspeople to start their own business as a building contractor.

Carpenters, Interior Systems. This work involves installing modern equipment and materials in commercial buildings: acoustical ceilings, raised floors for computers, metal framing or wall partitions, and office furniture systems, for example. These carpenters assemble complex interior systems using technical data supplied by manufacturers. Materials arrive marked and crated, requiring workers who can read and carefully carry out the instructions. As the term "interior" implies, most of this work is done indoors. Sometimes the work is at floor level, but workers may be required to use a platform or scaffolding.

Interior systems carpenters generally need carpentry tools such as saws and hammers as well as mechanic's tools such as drills, wrenches, and screwdrivers. They may also need welding skills for some of the assembly work. In this field, too, apprenticeship training is available. Wages follow the same patterns as for other carpenters.

Cabinetmaker or Millwright. These carpentry specialists display a high level of patience and precision in their work. They generally work in mills or factories, using modern tools and often assembly-line methods to cut, shape, and assemble wood products including moldings, panels, and furniture. They also produce high-quality fixtures for residential, office, and commercial use, including the cabinets and case goods interior systems that carpenters install. Such workers use fine materials and produce high-quality products, sometimes built to order in a specialty known as

The Specific Trades

architectural woodworking. These craftspeople operate machines including power saws, planers, joiners, and shapers. Apprenticeship training is generally available.

On a daily basis, architectural woodworkers read scale drawings and blueprints, select materials, make precise measurements, and cut, glue, and clamp elements of their project. They also assemble parts into a completed project and create a smooth finish that enhances the materials. Finishing details such as hardware are also installed.

Of course, the breakdown of duties varies from one workplace to another, usually dictated by the size of the operation. The smaller the shop, the more likely a worker is to control all or most elements of construction, from planning to finished product.

Persons best suited to such work pay close attention to detail and safety and take pride in their work. Apprentices generally start by helping journeyworkers with basic tasks such as material handling, sanding, and some assembly work, moving into more sophisticated jobs as their knowledge and skills increase. You can expect to learn about characteristics and properties of wood and other materials. Hourly wages in early 1993 averaged between $18.72 in the Southeast and $30.63 in the Pacific district.

Construction Inspectors

Construction inspectors examine the construction, repair, and alteration of building projects including structures, bridges, highways, and dams. They determine whether the projects meet required building codes and ordinances, contract specifications, and sometimes zoning regulations.

Inspections begin with the start of construction and continue periodically until the project is completed. At that time a rigorous final inspection is done before the building can be occupied or the structure put into service. Inspections must be made at the building site before the foundation is poured, since soil condition and the position and depth of footers and piers are of critical importance. Inspection is made again after the foundation is installed. In the case of highway inspections, this careerist is often the official who determines when the contractor can be paid for certain phases of the work.

At the close of the 1980s nearly 70,000 people were at work as construction and building inspectors, more than half of that number working for local governments. Employment is concentrated in cities and in suburban areas that are experiencing growth.

Computers have gained a foothold in this work, since it involves many details and a great deal of paperwork.

The inspector is well trained in construction theory, usually specializing in one type of structure or project. Much of the work is done through visual inspection, but it often involves use of tape measures, survey instruments, meters, and other test equipment. Many inspectors keep a daily photographic record of important aspects of a project. They generally work alone, although a large or complex project may be inspected by team effort.

Inspectors spend much of their time in fieldwork but also work from an office, examining blueprints, filing reports, handling phone calls and correspondence, and scheduling inspections. Should an inspection reveal a deficiency, the inspector notifies the contractor or job supervisor. Problems that are not corrected within a reasonable time may result in issuance of a stop-work order, particularly in cases of government projects.

People entering the inspection field generally have

acquired several years of practical experience in one of the building trades. For instance, a plumbing inspector has a journeyworker's background as a plumber. They must have a thorough knowledge of construction materials and accepted practices. Increasingly, inspectors prepare for their work in an apprenticeship or by acquiring a two-year degree in construction technology, combined with practical work experience in a related craft. Federal and state governments sponsor ongoing training seminars for inspectors, which are open to government employees and inspectors employed in private industry.

Inspectors usually work a 40-hour week with daylight hours. Inspection sites may be hazardous because of accumulated materials or debris. The inspector may be required to climb ladders or manuever in cramped quarters, although the job is not considered as physically demanding as most construction work.

As a result of more stringent training requirements, many states now require building and project inspectors to pass a certification exam. Those who are certified generally earn higher pay and enjoy increased job opportunities.

The outlook for this occupation is expected to keep pace with job growth in other occupations through the year 2000. In 1992 the median wage for construction inspectors was just over $30,000 a year, which means that half of all inspectors earned less than $30,000 and half earned more than that.

To learn more about career opportunities as a construction inspector, write to the American Society of Home Inspectors, Inc., 3299 K Street NW, Washington, DC 20007; or the International Association of Plumbing and Mechanical Inspectors, 20001 South Walnut Drive, Walnut, CA 91789.

Drywall Mechanics and Lathers

These workers construct and finish interior walls and ceilings. They make accurate measurements and cut materials with precision to cover walls and ceilings.

Most of these workers began as helpers, transporting materials and providing assistance to experienced workers by lifting and supporting panels as they are installed and by keeping the workplace clean. After demonstrating their willingness to work and learn, they began measuring, cutting, and installing materials. More recently, formal apprenticeships have become the recognized route to quality job training. Such programs generally last two years and blend classroom instruction with actual job experience.

The work of creating finished interior spaces begins when the carpenters have installed structural elements of the load-bearing walls as well as the interior partitions. Residential jobs use primarily large sheets of gypsum wallboard, sometimes called sheetrock, to cover the walls. Thus workers who install sheetrock are often known as "rockers." These craftspeople work quickly, often in teams of specialists who hang the dry wall, followed by those who tape its seams, and concluding with workers who finish the seams, creating a smooth surface for the final decorative treatment of paint or wall covering.

Lathers form a small but vital segment of the construction industry. They work with strips of metal or gypsum lath, preparing a framework on which plaster is applied to create walls or ceilings. Gypsum lath is similar to drywall panels but is cut in smaller sizes. Lathing is an ancient trade that dates back to the days when palaces and cathedrals were built from stone and plaster. The ornate ceilings and walls of many public buildings would be impossible without the skills of

The Specific Trades

the lather, who applies the basic framework for the plasterer.

Although lathers work indoors much of the time, they may also work from scaffolding on the outside of buildings and other structures, often at considerable distances from the ground. Lathers use more complicated tools than does the drywall mechanic, since metal lath frequently must be cut, joined, and fastened using welding equipment and oxyacetylene torches.

Drywall mechanics use razor-sharp utility knives to cut materials, power screwdrivers, and staple guns. For many years drywall tape was a simple paper material covered with a plasterlike compound. In recent years technological advances have brought a switch to stronger, more effective plastic and Fiberglas tapes and compounds that accommodate settling of buildings without causing cracks in the finished walls.

Drywall mechanics work indoors and face many of the hazards of the lather, since they often work from a scaffold. Those who finish ceilings may even use stilts as a means of combining height and easy mobility.

Both trades include strenuous work requiring dexterity to lift and manuever heavy panels. These are also jobs with constant motion since the worker stands, kneels, and bends for much of the workday.

Approximately 150,000 people do this work, most of them living near urban centers where the big construction jobs are concentrated. Those who specialize in residential construction may face layoffs during building slowdowns, and there may also be downtime between assignments.

Opportunities for advancement include becoming a job supervisor or launching one's own contracting business.

Employment opportunities for drywall mechanics and lathers are expected to keep pace with other construction trades through the year 2000. New job openings will arise, and workers will be needed to fill vacancies as workers retire, start their own business, or move on to other trades.

According to recent figures, union drywall mechanics and lathers earn between $14 and $29 per hour. Apprentices start at half a journeyworker's wages, moving up with experience. Most building contractors pay an hourly wage, although it is not uncommon to be paid according to the number of drywall panels installed or finished per day.

The standard workweek is 40 hours, with overtime available at the height of the construction season. Both the International Brotherhood of Painters and Allied Trades and the United Brotherhood of Carpenters and Joiners of America represent these crafts.

Electricians

Electricians spend their workdays installing and maintaining electrical systems and equipment. They are an essential part of any building crew; like carpenters, their work is divided into "rough" and "finish" categories.

At residential building sites electricians install electrical circuits, appliances, and such equipment as water heaters, smoke or burglar alarm systems, intercoms, and heating systems. Commercial construction involves power supplies at shopping centers, supermarkets, office buildings, and service stations or garages. A third category of construction electrician jobs is known as industrial or "large" work. These job sites might include new factories, petroleum or chemical plants, and other production facilities where electricity is required to heat, light, and power the operation.

All electricians are trained to follow National Electrical Code specifications as well as state and local electrical codes. As you might imagine, observing safety

The Specific Trades

precautions is essential in this career field, and safety education constitutes a major part of job training.

Whether the project is a suburban home or a high-rise office tower, the construction electricians follow blueprints and architectural specifications as they install wiring systems. Other job activities could include installing electrical machinery and controls and signal and communication systems. Because electronic controls are becoming increasingly important in home and commercial building design, construction electricians are involved with this new aspect of wiring as well.

Many construction electricians are employed by electrical contractors, of which there were nearly 50,000 in the U.S. in the late 1980s. The U.S. Department of Labor projects a need for nearly 6,500 construction electricians per year through the year 2000. Qualified electricians work anywhere in the country, although the largest concentration is found in industrialized and urban centers.

Electricians use a variety of tools on the job. Usually they supply their own hand tools including screwdrivers, pliers, hacksaws, and utility knives. Employers supply such items as pipe threaders, conduit benders, power tools, and test meters.

This career does not require great strength, although good physical condition is a must, since it requires standing for long periods, often in confined spaces. A minor yet vital matter is good color vision, since electrical wires are often color-coded.

Those who work in electrical crafts require a natural aptitude for using tools and the mental ability to understand the scientific principles of electricity. Most construction wiring occurs indoors. There is an element of danger; workplace injuries include falls, electric shock, and being struck by falling objects.

A formal apprenticeship is considered the best way to

become an electrician, so it is not surprising that the majority of today's electricians have spent four years in such training. In most states electricians must pass a licensing exam that tests knowledge of the craft and of building codes, both state and local.

Experienced construction electricians may be promoted to supervisory positions. Electricians are among the highest paid of all construction workers. Salaries ranged from $20 to $33.03 per hour in 1993. Fringe benefits make the package even more valuable. Because of the seasonal nature of construction work, however, such wages may not be available year-round.

More than a half million electricians are employed in the U.S., with just over half of that number in the construction industry. Fewer than 10 percent of all electricians are self-employed. The demand for qualified electricians is expected to keep pace with the growth of jobs in other segments of the economy through the turn of the century. New technologies such as the electronically controlled "Smart House" (see Chapter 11) are expected to add to this demand.

Whereas job turnover is prevalent in other construction trades, electricians do not change jobs so often, partly because of the length of training, and partly the relatively high pay. The International Brotherhood of Electrical Workers and businesses that employ electricians are making a concerted effort to recruit talented young men and women to fill the ranks; indeed, apprenticeship slots for electricians are second only to carpentry apprenticeships.

Floor Layers

Floor layers work in commercial buildings, hotels, homes, and churches. They install carpeting, hardwood flooring, and resilient flooring materials of sheet or tile in vinyl or rubber.

The Specific Trades

Floor layers cut, fit, and install these materials and various types of underlayment to insure smooth, level surfaces for the finished floor. They scribe, cut, lay out, and seam tile and "sheet stock" in a variety of patterns. They are also skilled in cutting, binding, sewing, and installing carpet. They work from the specifications of architects and interior designers and must be good at estimating materials. The best floor layers can visualize and plan intricate geometric designs, often including installation of inlaid pieces.

Most of the work is performed indoors. Floor coverings are installed on all types of surfaces, so floor layers must know the safe and proper use of many types of adhesives and fastening systems.

Because floor layers are called upon to replace old flooring in furnished buildings, they must be neat and careful to avoid damaging existing fixtures or furnishings. Likewise, floor coverings may be the final element in a new building, requiring the workers to be careful of the work of craftspeople who have gone before them.

Floor-layer training is available in many cities through union apprenticeship programs. This combination of on-the-job and classroom training takes three to four years.

Nearly 60,000 people work as floor-covering installers, many employed by flooring contractors or retail stores that sell floor coverings. Many of them learned their trade informally, starting as helpers. Formal apprenticeships exist, lasting up to three years and combining class time with on-the-job training.

Opportunities to advance include becoming a supervisor, job estimator, or sales manager for a floor-covering sales operation. Experienced carpet installers can secure a measure of independence by working as subcontractors for building contractors.

33

Apprentices learning to install floor coverings work with carpet, vinyl, wood, and other modern materials (courtesy International Brotherhood of Painters and Allied Trades).

The combination of new construction and remodeling makes the outlook bright for floor-covering installers. Information on earnings is limited, but a range from $25,000 to $50,000 per year has been cited in several government studies. Most floor-covering installers are paid by the hour; some receive a monthly salary or are paid according to the amount of carpet installed per month. Unionized floor-covering installers generally are represented by the same craft union that represents painters and glaziers.

The Specific Trades

Glaziers

Glaziers are skilled workers who select, cut, install, and remove glass of all types, as well as synthetic substitutes. Because glass is strong, visually attractive, and increasingly energy-efficient, it has evolved as an important design element in structures of all types. Windows and glass walls are not the only projects with which glaziers are concerned, however. They also install mirrors, shower doors, display cases, glass doors, and other architectural details.

For some jobs the glazier must cut glass stock to size at the job site, securing it in a frame. They use tools including glass cutters, suction cups, glazing knives, power saws, drills, and grinders.

This work is often physically demanding, done both indoors and out, sometimes in inclement weather. The necessity for working from ladders or scaffolding adds an element of danger. Glaziers risk injury from cutting tools, broken glass, falls, or falling objects at a construction site.

Nearly 50,000 persons follow the glazier's trade, most of them employed by specialty contractors or construction firms. Others work for retail concerns that install or replace glass products including windows and doors.

Today an apprenticeship program lasting three to four years is generally accepted as the best way to learn this trade. Advancement is somewhat limited, although some glaziers become estimators or start their own contracting firms.

Demand for this skill will keep pace with most other occupations through the year 2000. Growth in commercial and industrial construction is an important consideration, however. As is the case with several other building trades, glaziers may experience periods of unemployment when construction is in a cycle of decline. Anticipating a shortage of qualified workers, employers and unions have begun actively recruiting the next generation of glass installers.

Information on pay scales is sketchy, but hourly wages depend on geographic area. Union journeyworkers reportedly receive from $18 to $24 per hour. Weather conditions and downtime between construction projects can also affect total earnings.

Heating, Ventilation and Air-Conditioning (HVAC) Mechanics

These workers install, maintain, and repair the increasingly technical machinery and equipment used to warm, cool, and circulate the air in homes and buildings. Usually, these mechanics specialize in one or two areas of the field. Surveys show that one in seven is self-employed.

The size of the equipment the HVAC worker deals with can vary widely, from room air conditioners to room-size units capable of cooling an office tower. Industrial applications may involve even larger units; for example, refrigeration units for a frozen-food processing plant. Regardless of equipment size, the basics of installation and maintenance vary little. Refrigeration mechanics are trained to read blueprints and manufacturer's specifications. They also install refrigerant lines and piping and connect ductwork and electrical power.

Heating mechanics install furnaces, steam boilers, and heat pumps, working with fuel oil or natural gas-burning equipment. Such mechanics are also trained to maintain and repair this equipment.

Jobs in air conditioning and refrigeration are available across the U.S. and Canada. Some heating mechanics specialize in oil burners, others in natural gas-fired equipment. Many work for public utility companies and plumbing and heating firms. About 20 percent of such workers are self-employed.

The Specific Trades

A 40-hour workweek is typical except during peak heating or cooling seasons, when breakdowns may bring repair workers overtime hours.

The work may involve an element of danger, often requiring working at some height and in cramped or awkward positions. Other hazards include the possibility of electrical shock, burns, or injury from lifting heavy objects. As you might expect, strength, agility, and excellent physical condition are required.

The typical HVAC worker has devoted two years of training beyond high school to learning the craft. Vocational, technical, and community colleges offer training programs. Some HVAC mechanics receive on-the-job training through formal apprenticeship programs lasting four to five years and combining hands-on work with classroom instruction.

About 180,000 HVAC mechanics are already at work in this field, but employment opportunities are expected to increase rapidly, especially for air-conditioning and refrigeration mechanics. Those trained to install, repair, and maintain heating equipment will also be in demand, although their opportunities are more closely linked to supply and demand factors in the construction industry.

Earnings range from $15 to $19 per hour, with those versatile enough to work on several types of equipment able to earn considerably more. Helpers and apprentices earn between 40 and 80 percent of the fully trained worker's pay.

Insulation Workers

These construction workers lead the way to the future by creating energy-efficient buildings. Insulating materials may be solid—including Fiberglas batts and loose-fill—or rigid foam in panels or in tubular shapes to insulate pipes. Insulation may also take the form of

The heating, cooling, and ventilation contract for an office building of this size requires the skills of dozens of HVAC workers. Here a section of metal ductwork is being readied for installation (courtesy Home Builders Institute).

sprayed-on materials that quickly harden through a chemical process.

Keeping warm air from transferring to a colder atmosphere is the basic goal of most residential insulation. Insulation workers use common hand tools including stapling guns, power saws, and air compressors. Most of the work is done indoors and involves a great deal of bending, kneeling, and standing. Insulators sometimes work from ladders and in cramped spaces. Coordination is more important than physical strength. Workers must wear protective clothing and guard against inhaling airborne particles and irritants from the materials they work with.

The Specific Trades

About 65,000 insulation workers held jobs in the U.S. in 1990. On-the-job training and formal apprenticeships are the most common ways of entering the field. The primary union is the International Association of Heat and Frost Insulators and Asbestos Workers. Job advancement consists of promotion to supervisor, shop superintendent, or contract estimator, or launching a business of one's own.

Growth in numbers of these craftspeople will keep pace with the average for all occupations through the year 2,000. New buildings must be insulated, and existing buildings will be renovated and brought up to modern standards of insulation.

In 1993 the average hourly wage for insulation workers was listed as $29.77 including fringe benefits. Workers who specialize in commercial and industrial work earn more than residential insulators, since the latter do not require as much skill.

Laborers

Laborers account for more than a dozen job categories in the construction industry. These are the people who perform the physical labor that cannot be accomplished by machinery. They may also run machines including power lifters, small mechanical hoists, and motorized ditch diggers.

Still, they perform tough physical work. In a typical workday they might erect and dismantle staging and scaffolding; mix, pour, and smooth concrete; unload and move stacks of building materials; keep the work site free of debris; install and connect drainage pipe in trenches, and grade and prepare highway right-of-ways.

This is one of the few jobs you can get without serving an apprenticeship. Construction laborers acquire training through rigorous on-the-job instruction. Some unions and contractors have put together four- to eight-week training courses, teaching safety practices, machinery operations, and basic construction theory. With experience, laborers who have demonstrated skill and the willingness to learn can usually move from the labor pool into a more skilled job with higher pay.

As of 1993 union construction laborers averaged $20.16 per hour, with fringe benefits worth 11.4 percent of their salary. Remember that these wages are for workers with no formal job training. Remember, too, that geographic location has a big impact on wage rates.

Through the year 2000, approximately 50,000 construction laborers will be hired each year. Currently more than 850,000 are at work across the nation. As with all construction jobs, the state of the economy affects hiring and job security.

The primary labor union of these workers is the Laborers' International Union of North America, 905 16th Street NW, Washington, DC 20006.

Land Surveyors

Surveyors establish official boundaries for land and water; write descriptions of land for deeds, leases, and other legal documents; and measure construction sites. They are assisted by survey technicians, who operate surveying instruments and collect information. Some surveyors collect information for use in maps and charts. It is their job to measure distances, directions, and angles between geographic points and determine elevations of points and contours of the earth's surface.

Surveyors also plan fieldwork and determine the precise location of important features of the area being surveyed. They may research legal records and look for evidence of previous boundaries. They record results of surveys, increasingly using computers to store data. They also verify the accuracy of data and prepare plats,

The Specific Trades

maps, and reports. Surveyors who establish official boundaries must be licensed by the state in which they work.

Sophisticated electronic equipment including lasers and satellite-guided instruments are among the space-age tools surveyors use. One aspect of the job that is not changing so rapidly is fieldwork. Surveyors spend much of their time out-doors inspecting the land to be measured. They usually lead a survey party in the day-to-day work, supervising survey technicians and helpers. Survey technicians adjust and operate the surveying instruments such as the theodolite, used to measure horizontal and vertical angles, and electronic distance-measuring equipment. Technicians or helpers hold the vertical rods the theodolite operator sights on to measure angles, distances, or elevations. This work requires participants to read and give proper hand signals and to read survey instruments. If the surveyor is not using electronic equipment to measure distances, the technicians or helpers may hold measuring tapes and chains to accomplish the same tasks. Survey technicians are called upon to take notes, make sketches, and, increasingly, enter the data collected in the field into a computer. Depending on the terrain, laborers may be needed to clear brush from sight lines, carry equipment, pound stakes, and perform other less skilled tasks.

Surveyors involved with construction projects generally work for contractors, state highway departments, public works agencies, or utility companies. Of more than 100,000 trained surveyors in the U.S., only about 10,000 are self-employed.

The workweek is basically 40 hours but may include overtime when the weather favors fieldwork. This is active work and may be strenuous if the terrain is rugged, for there is heavy equipment to transport.

In the past, community colleges and technical schools trained most survey technicians and surveyors in programs that lasted from one to three years. Typical programs combined on-the-job training with classroom study. The Job Corps also offers a training program for survey assistants.

All fifty states now require surveyors to be licensed, and an increasing number require a bachelor's degree. Those in the industry predict that all states will follow this route by the year 2000.

Employment opportunities for surveyors are expected to keep pace with job growth in all occupations through this decade. Anticipated growth in construction will create job opportunities for surveyors who lay out streets, housing developments, shopping malls, airports, roads, and highways. Still, the economic shifts that affect construction employment in general will be felt in this allied field.

Surveyors earned between $21,000 and $35,000 per year in 1990, depending on geographic location, experience, and training. Among the highest-paid were those in regions where the building season is longest.

Landscape Architects

These specialists prepare detailed maps and plans to show all existing and planned features. Once plans are approved, they solicit bids from landscape contractors for the work to be done. While a job is under way, the landscape architect supervises any necessary grading, construction, and planting. Placement of trees, shrubs, and walkways must follow the approved plan.

Most landscape architects are self-employed or work for a landscaping or engineering firm. State and local governments employ landscape architects to plan parks and recreation areas and work on highway design and urban renewal projects. A four-year college degree is

The Specific Trades

required, and passing of a state licensing exam in most cases. A less technical career in landscaping is available with only two years of preparation beyond high school. Working in a plant nursery or for a landscaping contractor is good experience, regardless of the level of achievement that is your career goal.

The entry-level landscape architect performs such duties as simple drafting assignments and works up to preparing project specifications and construction details. After two or three years the architect is given responsibility to see projects through to completion.

Work is available throughout the U.S., with the most growth in recent years in the Southwest. The workweek is generally 40 hours, although the self-employed often report working much longer hours. The landscape architect combines outdoor site work with office hours, whereas the landscape contractor's work is primarily outdoors and may also include operating a plant nursery.

About forty colleges offer a bachelor of science degree in landscape architecture that has been approved by the American Society of Landscape Architects. Sixty other schools offer programs or courses in the field. Some states permit the substitution of six to eight years of apprenticeship with a landscape architect for the college degree.

About 15,000 registered landscape architects are at work in the U.S. The outlook is for rapid growth, although it is strongly tied to that of the construction industry.

Most jobs are located near population centers. Advancement is generally achieved by moving to a larger firm or opening one's own business.

Beginning salaries in 1993 were in the low $20,000 range, particularly for landscape contract workers. Registered landscape architects earned an average of

$44,000, with the highest pay (approximately $54,700) found in private practice along the Pacific coast.

Operating Engineers

Also known as construction machine operators, these people work for contractors building housing developments, highways, and other large-scale projects. They also work in building construction, landscaping, grading, and excavating, or in hoisting materials. Others work for public utilities, manufacturers, and other firms that do their own construction work.

Operating engineers run power shovels, graders, bulldozers, trench excavators, tractors, pile drivers, dump trucks, backhoes, and motorized cranes. Some of the equipment is not so large, yet is indispensable on the modern building site, including hilifts, fork lifts, and lumber carriers; electrically propelled equipment; concrete, stone, and asphalt spreaders, and specialty paving equipment.

They may also operate nonmoving equipment such as pumps, gunnite machines, electric generating plants, refrigeration equipment, turbines, compressors, and diesel and steam engines.

On any construction site you are likely to find a wide variety of machinery ranging from the huge and complex to the small and simple. Most is designed to move materials, whether hoisting lumber or steel, digging a trench for utility lines, or transporting cement. In general, the larger the machine, the more complex it is to operate. Running heavy machinery such as a crane requires skills developed through years of practice. The operator needs to judge distance and height with pinpoint accuracy. To manuever the machine requires pushing and pulling many buttons, levers, and pedals in the proper order. The operator picks up and delivers materials to great heights, some-

The Specific Trades

times to locations that are out of sight. However, operating a bulldozer is similar to driving a car. Buttons or levers raise or lower a blade attached to the vehicle's front end. Until you acquire experience, you are assigned to the lightest equipment that requires the least skill to operate. A compresssor operator starts the engines and monitors fuel, water, and pressure levels.

There is a growing need for skilled people to operate equipment in construction. More of the functions that were once performed by unskilled labor have been taken over by machinery.

Dirt and noise are all part of a day's work for an operating engineer. Modern noise-limiting devices protect a worker's hearing and help to assure job safety by dampening the sound of roaring engines while still permitting the wearer to hear verbal instructions. The workday includes constant movement from the machine, along with work of an active outdoor nature.

The median annual earnings for all operating engineers in the mid-1990s was about $415 a week. Approximately 93,000 operating engineers are at work in the U.S.

For many years operating engineers were likely to receive their job training during military service. Most of today's operating engineers take part in a formal apprenticeship program operated by employers, locals of the International Union of Operating Engineers, or the U.S. Department of Labor. The Job Corps also offers a preapprenticeship program for people who seek apprenticeship slots.

To qualify as an apprentice, young men or women must be at least 18 and not over 25 years of age. They must satisfy the local Joint Apprenticeship Committee that they have the ability and aptitude to master the trade. A high school diploma or equivalent is required to complete the related theoretical instruction. During the selection process applicants are expected to demonstrate manual dexterity, mechanical aptitude, and excellent health.

The Joint Apprenticeship Committees have the authority to waive the maximum age limit and to give special consideration to eligible veterans and other special cases.

Apprentices serve a maximum of three years, or about 6,000 hours, reflecting six six-month periods of reasonably continuous employment (taking into account weather delays and downtime between assignments). This work experience is in addition to time spent in related instruction. An apprentice may be raised to journeyworker status after serving three years and passing an exam. Typically, wages start at about 70 percent of the rate paid to experienced workers, somewhat higher than many other apprentice wage scales. You can expect about 144 hours of classroom instruction each year, plus on-the-job training supervised by experienced operating engineers. Most likely, you will start by cleaning, greasing, repairing, and starting machines. Within the year you'll be practicing some simple machine operations. Over the three years, your responsibilities and skills will grow.

After your training you are eligible to join the union, but that step is not cheap. Initiation fees range up to $1,000. After that you pay regular union dues, but remember that you will receive the higher union wage rate. To get more information on the program, ask the

Apprentice carpenters nail up furring strips, to which sheets of gypsum drywall will be attached, thus creating a new ceiling. Instruction is provided by a journeyworker carpenter (courtesy Home Builders Institute).

local Union of Operating Engineers where to take the admission test for apprentices.

As construction work picks up, there will be a strong demand for operating engineers. According to 1990 employment figures, 820,000 people were employed in this trade, an increase of 41.4 percent in the past five years.

Painters and Paperhangers
These workers apply finishes that protect and beautify wall surfaces. They use a number of products and techniques on the job. Work environments include structures both old and new. Some people do both painting and installing wall coverings, although the jobs require different skills. Choosing and applying appropriate paint, stain, varnish, or other liquid finishes to buildings is a job that dates back to Colonial America. Surfaces must be prepared to receive the finish. Paperhangers also prepare surfaces to receive wall coverings of paper, vinyl, fabric, and other materials.

The work requires skill in measuring and in matching colors and patterns in addition to cutting and manipulating the material for a professional look. Painters and paperhangers often work from ladders or scaffolding. They must be agile and have no fear of height. These workers stand, bend, and climb throughout the workday. Also, much of the work is done with arms raised overhead. Falls are a common hazard. Adequate ventilation at the workplace is an important safety factor. Painters often work outdoors, and high temperatures may be a consideration during part of the working season. Cold, wet weather generally brings work to a halt.

Approximately 572,000 painters and decorators or wall-covering installers were at work in the United

Paint not only decorates but protects the surfaces of a building exposed to the weather. Here two young women are at work on a residential project (courtesy IBPAT).

States during 1990. Many are employed by contractors whose work is in new construction, repair, and restoration or remodeling. Organizations that own and maintain large buildings such as hotels and office and apartment complexes often hire workers year-round to carry out scheduled maintenance. Some painting and decorating firms serve institutional clients such as schools and hospitals. Industrial painting firms specialize in painting factories, bridges, even church steeples.

Both informal and formal training programs prepare

these workers for their careers. The three-year apprenticeship program includes 144 hours of classroom instruction on such topics as color harmony, surface preparation, material application techniques, use and care of tools and equipment, blueprint reading, cost estimating, and job safety.

Employment opportunities for painters are expected to keep pace with other construction job growth for the balance of the decade, with approximately 27,000 job openings being created each year by retirement of older workers and those moving on to other types of work. The field for wall-covering installers is much smaller, with growth of only about 1,500 jobs each year. Many of the workers in both of these categories are self-employed. Because new construction jobs are generally of short duration, most of these workers fill in their schedules by contracting short-term jobs with small businesses or the owners of private residences. The ability to market one's skills is essential to successful self-employment.

Like other construction workers, painters and wall-covering installers have been making pay gains over the past several years. Median wages for union members were $382 per week. The top 10 percent of all wages went to maintenance painters, who averaged $677 per week. Fringe benefits added to the package. Supervisors can earn about 10 percent more, and the highest wage trends occur in California. Some custom firms report earnings up to $1,000 per week in cities such as New York, where interior designers seek craftspeople who can supply artistic or specialized wall treatments.

In general, interior decorators or wall-covering installers earn more than painters. Painters who do primarily maintenance earn less than other construction painters, yet their salary is still higher than that paid to

THE SPECIFIC TRADES

nonsupervisory production workers in other segments of the economy.

Many painters are nonunion, although thousands belong to the International Brotherhood of Painters and Allied Trades.

Plumbers

Plumbers are in demand at construction sites, and their compensation shows it. A survey of skilled workers in the construction industry found plumbers earning from $17.86 to $45.78 per hour during the summer of 1989. By early 1991 the U.S. average was $30.15 per hour, exclusive of fringe benefits. That figure had increased to $32.99 by January 1993.

If to you the plumber is the person who comes to your house to replace a leaking pipe or install an appliance, you may be interested to learn more about the role of plumbing and pipefitting in construction.

It is the construction plumber's job to install and repair the piping systems that carry water, waste, drainage, and natural gas in all sorts of buildings. Installing plumbing fixtures such as bathtubs, sinks, and toilets is also the plumber's job, along with installation of appliances including dishwashers and water heaters.

Plumbers work from blueprints or drawings that indicate the location of pipes, plumbing fixtures, and appliances. They lay out pipelines in such a way as to conserve materials, but they must meet stringent inspection requirements depending on the type of building and the materials used, from rugged plastic pipe to copper or cast iron. Sometimes they must cut holes in floors, walls, or ceilings. They may also install steel pipe supports from ceilings. These activities, plus cutting, bending, and assembling lengths of pipe, require the use of various hand and power tools. Some pipe connections are made by soldering, which requires the use of a torch. The final step in finishing a piping job is to test the system for leaks, using pressure gauges. When that is done, fixtures required for the building may be installed.

Plumbers need physical stamina to carry out a job that often includes lifting and carrying heavy pipes. They often stand, kneel, or crouch for long periods of time. It may be necessary to work outdoors in inclement weather. Workplace hazards include the potential for falls, burns from hot pipes and soldering equipment, and cuts from sharp tools.

Construction plumbers generally work 40 hours per week, with seasonal opportunity for overtime. Nearly 400,000 plumbers and pipefitters were at work in the U.S. in 1993, about two thirds of them working in new construction. One of every five plumbers is self-employed.

Apprenticeship programs, both union and nonunion, form the backbone of preparation for this career. Four years of on-the-job training is combined with classroom instruction. In addition to learning to read blueprints and know local plumbing codes and regulations, apprentices study job safety, construction math, some physics, and chemistry. At the job site they learn to identify grades and types of pipe, to move materials safely and efficiently, and to handle the tools of their trade. With experience, the apprentice moves on to installing various types of pipe.

Although there is no uniform national licensing program, nearly every community requires that plumbers be licensed by examination. These jobs are less sensitive to downturns in the economy than many other construction trades. When new construction slows, plumbers usually turn to maintenance or

THE SPECIFIC TRADES

rehabilitation projects, but some downtime may occur between jobs.

Advancement opportunities include becoming a supervisor for a mechanical or plumbing contractor. Many experienced plumbers open their own contracting businesses.

Roofers

Roofers install and repair the covering that protects the interior of a structure from the weather. Regardless of size or purpose, every structure requires a roof, and it is the roofer who installs and repairs these systems. A wide variety of materials and of installation procedures are involved in the business. Basically, roofs are of two types: flat and pitched (slanted). Most houses have pitched roofs, whereas office buildings, factories, and commercial properties have flat roofs. Regardless of the materials applied, the goal is a watertight surface. In recent years flat roofs have benefited from new technology and materials, including waterproof rubber and thermoplastics, materials that are safer and easier to work with, long-lasting, and durable.

Pitched roofs are generally covered by shingles or tiles. This work requires cutting and fitting materials around intersecting rooflines, chimneys, and vents. Power nailers are commonly used, and mechanical lifts assist in delivering heavy material loads.

Roofers still work under hazardous conditions, and, needless to say, at greater heights than many of their coworkers in the construction trades. This is demanding physical labor, requiring lifting and carrying of heavy materials. A roofer's day is filled with climbing, kneeling, and bending. Knowledge of tools and equipment is necessary to ensure a safe work area. Even so, roofers have one of the highest rates of job-related injuries. Work proceeds in all kinds of weather, particularly summer months when rooftop temperatures can exceed 100 degrees F.

It is relatively easy to learn the roofing trade without formal instruction or apprenticeship, although such programs are available. Many roofers begin as material handlers or roofer's helpers, learning the trade alongside an experienced worker.

Because roofs are exposed to extremes of weather, they deteriorate faster than other structural elements of a building. Roofers are in demand for replacement as well as new construction. In fact, 70 percent of the roofing business is in replacement. The field also has a high turnover rate as roofers learn other construction trades and move into new job openings. Spring and summer are the busiest times of year, although in some climates roofing businesses flourish year-round.

In 1993 the median weekly earnings for roofers in the U.S. were about $360. The top 10 percent earned $600 or more per week.

Structural Steel Workers and Ironworkers

These workers are employed by general contractors, bridge and dam construction companies, public utilities, and government agencies. They erect, align, and fasten the metal framework, whether structural iron or steel, that supports bridges, buildings, storage tanks, and other structures.

Cranes are used to hoist structural beams or girders to the work area, where ironworkers are positioned to push, pull, and pry the beams into place. After bolting the material into place temporarily, the worker measures for correct alignment before permanently fastening it with bolts, rivets, or welds.

Riggers and machine movers are ironworkers who prepare the hoisting equipment used to erect or dismantle structural steel frames. They are also responsible

The Specific Trades

for moving heavy machinery around the construction site. Both jobs require thorough knowledge of hoisting equipment and devices.

Reinforcing ironworkers represent yet another specialty within this craft. They work closely with concrete crews, putting steel rods, bars, or coarse mesh into forms to reinforce concrete as it cures.

All four of these branches of ironworking are expected to provide a good number of jobs throughout the 1990s, with employment expected to increase much faster than the average for all occupations. More structural workers will be needed because of the growing use of structural steel in buildings. Riggers and machine movers will be required to handle ever-increasing amounts of heavy construction machinery, and the greater use of prestressed concrete will call for more reinforcing ironworkers.

Still, the job has disadvantages. Avoiding injury is a prime concern, for a construction site can be a dangerous place. The development of safety nets and work harnesses has been a great help to workers who risk falls from extreme heights. Increased emphasis on keeping work sites clear of debris and on safer material handling techniques has also cut the number of injuries from falling objects.

Ironworkers are strong, for this is not light work. Manual dexterity, excellent eye-hand coordination, and the ability to work at considerable heights are all characteristic of these careerists.

As in most kinds of construction work, the time spent on the job is subject to changes in the economy and the weather. The total number of hours worked is less than that in private industry, but the hourly wage is about twice that of a nonsupervisory blue-collar worker. The industry is highly unionized, so those with the least seniority are furloughed first. However, the union provides supplemental unemployment benefits that make unemployment checks more bearable.

Most of these construction workers are members of the International Association of Bridge, Structural and Ornamental Iron Workers. Because of the obvious danger, high steel workers earn more than their colleagues who specialize in ornamental ironwork; however, most wage tables do not separate the job categories. Nonunion workers earned a median wage of $29.79 per hour in 1993, plus fringe benefits. Wage rates vary considerably from one part of the U.S. to another, with the lowest rates generally reported in the South Central states. The highest wages were reported in the Middle Atlantic and Southwest regions. Union wages followed the same geographic distribution, ranging from a high of $34.38 per hour in the Middle Atlantic states to a low of about $18 per hour in the South Central region. Comparable gains in fringe benefits accompanied the union-negotiated contracts.

As for training, apprenticeship is definitely the way to go. From the start, you are tested for agility and a good sense of balance. Most apprenticeships are sponsored by union-contractor agreements and last three years, during which you work under the supervision of an experienced journeyperson. You also attend 144 hours of classroom instruction each year, learning to read blueprints, use various tools, perform layout work, and acquire skill in welding. You learn all four types of ironworking, including ornamental, which requires you to install stairways, window frames, floor gratings, ladders, and fences. This work involves assembling and fitting ornamental iron pieces that have been created by other workers. Apprenticeship openings occur most often in early spring, just before the start of the building season, which means that you must start the application process months earlier.

The Specific Trades

Tile, Marble, Terrazzo, and Dimensional Stone Installers
This craft dates back to ancient Egypt and Rome. These workers are highly respected by their fellow craftspeople, since the work approaches the level of true art. Tile and marble setters are in great demand in the Southwest and West Coast regions particularly, where geometric designs are often integral elements in building facades.

Commercial buildings, offices, and upscale residential homes offer plenty of opportunities for tile setters to ply their trade. Tile is a favorite building material because it is waterproof, durable, and easy to clean. Tile, marble, and stone are applied to both floor and wall surfaces. Terrazzo is primarily a floor covering. In each case a smooth, flat surface is prepared to receive cement or mastic, which is applied with a trowel. The tools of these trades have changed little over the centuries. Attention to detail, including reproducing intricate designs from blueprints or drawings, is often required. Because natural materials vary in color and design, tile setters are often required to lay out material patterns on a dry floor before the actual installation. When the materials have been set in cement or mastic, tile setters fill the joints with grout, a fine cement that is often tinted to enhance the tile, and scrape away the excess before the grout sets.

The work is strenuous, requiring workers to bend, kneel, and reach. To protect their knees, many tile setters wear kneepads. The skills required are similar to those employed by bricklayers and stonemasons.

Tile, marble, and terrazzo workers work both indoors and outdoors. They create floors for patios and the intricate designs of tile in swimming pools. Union installers do most of their work in commercial buildings such as shopping malls, banks, schools, hotels, and apartment complexes. Their work includes floors, tubs and showers, countertops, and wainscoting. They mix mortar and install metal mesh to hold the mortar in place for tile setting. Tools of the trade include tile-cutting devices, mortar mixing equipment, heavy stapling guns, and tools to measure and level.

Information on salaries in these careers is limited; however, hourly rates ranged from $11 to $20 in 1989. Wages vary greatly from one part of the country to another. In 1990 just under 30,000 tile setters were at work in the U.S. About one in four is self-employed. Some operate their own shops, where customers select materials to be installed in custom-built homes or remodeling projects. Others work as subcontractors to building contractors.

The primary unions involved with the work of tile setters include the International Union of Bricklayers and Allied Craftsmen International and the United Brotherhood of Carpenters and Joiners (Tile, Marble and Terrazzo Finishers Division). Union apprenticeships are available in major cities in the U.S.

3

Union or Independent Craftsworker?

In parts of the U.S. and Canada, most or all of the people working in the building trades are members of labor unions. These organizations are called craft unions because membership is generally limited to persons who have many of the same skills and perform similar jobs. For example, the International Brotherhood of Carpenters and Joiners of America is the primary union for carpenters; the International Union of Operating Engineers represents skilled workers who run heavy equipment, and the International Union of Bricklayers . . . ah, you get the picture. Although there is some overlap, most craft unions are rather clearly defined in membership and function.

The purpose of a labor union is to bargain with contractors on behalf of union members, reaching industry-wide agreements on important issues such as pay scales and working conditions. For example, the union movement was instrumental in establishing the 40-hour workweek as a norm.

WORKPLACE SAFETY

The craft unions have also contributed greatly to improving the safety of workers engaged in construction. For example, in the early days of the International Brotherhood of Electrical Workers (IBEW), the work was so hazardous that apprentices often died in the attempt to learn the trade. In this fledgling yet dangerous new industry, safety standards were lax or nonexistent.

According to an IBEW report, during the late 1800s the national death rate for electrical workers was twice that for workers in other industries. When insurance companies learned what these people did for a living they refused to cover them, creating an even more precarious situation for their families. One of the first actions of the IBEW was to establish a $50 death benefit for every member. At that time the workday was typically 12 hours, often in all kinds of weather. The pay was 15 to 20 cents per hour.

Not every craft union has such a dramatic history, but all share the IBEW's concern for a safe workplace and adequate training for newcomers. The craft unions were among the earliest supporters of formal apprenticeship. The federal Job Corps program, with its construction skills training, also received early and enthusiastic support from the construction trades unions.

Today's craft unions are concerned with social service programs, whether offering members affordable insurance plans or volunteering time and talent to community projects.

PAYING YOUR DUES

In return for union representation and services, members pay a one-time initiation fee and monthly dues. Dues cover the costs of running the organization including union representatives, attorneys, research workers, and an administrative staff. Each union local has officers elected by the membership and draws up bylaws reflecting grass-roots issues. Members usually

Union or Independent Craftsworker?

elect one or more business agents whose job it is to run the union hall, deal with management on a daily basis, locate work for members, and assign them to jobs.

To be eligible for membership benefits, a worker must keep dues paid up. A union member always carries his or her dues book or card to confirm membership in good standing.

Some areas of the country require union membership; in others, construction workers have a choice of union or nonunion employment. It is up to the individual to determine the advantages of union representation, but statistics confirm that union members earn significantly more over a lifetime of work than nonunion construction workers. Union workers actually put in fewer hours on the job since they receive paid vacations and more paid holidays. In inflationary times union members are protected by wage contracts calling for COLA, cost-of-living adjustments, which automatically raise wages if the rate of inflation reaches a specified limit.

There are seventeen building trades unions in the American Federation of Labor–Congress of Industrial Organizations (AFL–CIO). Most of these workers are employed in commercial or heavy construction, along with the service industry. Less than half of all construction workers in the home-building industry are union members, probably because the scattered nature of the jobs makes it difficult to organize workers and conduct elections.

To a certain degree, union construction workers are shielded from the seasonal nature of their jobs. Higher overall wages make it possible to cushion an economic downturn or extended bad weather if the worker is prudent enough to put spare cash aside. A number of unions provide "sub-pay" supplementing members' unemployment benefits during the off-season.

Joining a Union

Most of today's union workers "graduated" to membership after completing a union-sponsored apprenticeship of three or four years. Craftsworkers who follow this route must pass a test given by the union local and pay an initiation fee, which may be up to $1,000. Those who have become journeyworkers by some method other than apprenticeship may also seek union membership by taking the admission skills test and paying the initiation fee.

Remember that unions constantly work toward job security for members. Seeking to maintain a balance between the number of available jobs and the pool of available workers is an important union issue. The number of apprenticeship slots is affected by this supply/demand issue.

4

Opportunities for Women and Minorities

The decade ahead promises to be an exciting time for women and minority workers who seek a career in the construction industry. As contractors face the reality of a changing workforce, people with good basic skills are being tapped for the best apprenticeship slots, and these candidates are increasingly females or members of minority groups.

Experts in employment trends forecast that one of every three construction job offers will be made to a minority or female worker by the end of this decade. The trend has been documented by steady growth in the number of apprenticeship slots awarded to women and minorities, beginning in 1978 when Title 29 of the Equal Employment in Apprenticeship legislation was adopted. In the year before that, fewer than one in six apprentices was a minority member, and fewer than 2.5 percent were female.

By 1983 minority placements exceeded 20 percent and women logged nearly a 4 percent increase. In 1990, 283,352 apprentices were enrolled in training programs across the U.S., 22.5 percent of them from minority groups and 7.1 percent female.

What construction careers are these people choosing?

Apprentice glaziers demonstrate their skill in removing a broken pane of glass (courtesy Home Builders Institute).

Those that pay well and command the most respect, according to industry surveys. Electrician apprenticeships attracted 1,646 females and 4,951 minority candidates; 1,411 females and 4,608 minorities entered carpentry apprenticeships.

Record numbers of females and minorities are entering the building trades, and they must continue to do so if a sufficient supply of skilled workers is to be trained for the future. Training programs including the federal Job Corps actively recruit these people and provide positive role models. For example, in the Job Corps female instructors teach the plastering trades and demonstrate the operation of heavy construction equipment. More about that later in this chapter.

Opportunities for Women and Minorities

The combined weight of several factors will keep the employment gates open and ever widening. Government supervision of industry hiring practices is a major influence. Contractors who bid on government contracts are already working with target figures in hiring. Enlightened attitudes on the part of unions and employers are taking effect, and the explosion of myths about gender and the nature of work are having a significant impact. As the construction trades compete with other industries for the best and brightest among a new generation of workers, salary increases and more attractive working conditions will result.

Many of the newcomers to construction in the last decade have already left the ranks of employees to become contractors. Their attitudes on hiring are bound to influence other employers.

Equal opportunity is a cornerstone of government-funded training programs such as the Job Corps. Both federal and state agencies work closely with employers, unions, and vocational schools to improve the quality and availability of apprenticeship training. Many adult literacy programs also offer job training programs with an emphasis on nontraditional careers.

Why do women seek jobs in the construction industry? For most of the same reasons men do: They are looking for interesting work that pays well. Many of the women who enter apprenticeships say that a family member—usually a father, a brother, or an uncle—was their role model. Others simply prefer an outdoor job to being cooped up in an office. And the economic advantages cannot be denied. Even training wages in construction jobs exceed the pay rate for so-called women's occupations in business or the service industry. By the end of training a woman in a blue-collar job can expect to earn four or five times as much as if she had stayed in a traditional job.

Making History

The summer of 1972 was a breakthrough season for women interested in entering the field. That was the year Chicago recorded its first female carpentry apprentice, a 28-year-old woman who spent her first day on the job installing cupboard handles and toilet-paper holders in a new apartment building. That same summer Carpenters Local 198 in Dallas accepted its first female member.

In early 1974, acting on results of interest surveys, the Job Corps established its first coeducational program to train workers for union construction trades. The program had the full support of the Building Trades Union, and from its earliest days it maintained a reputation for students who demonstrated sincere motivation and willingness to learn nontraditional skills, including those demanding endurance.

Recruitment Programs

Now that more than two million young people have successfully completed Job Corps training, active recruitment programs to place women and minorities in nontraditional programs continue to be a priority. Job Corps centers across the U.S. currently offer programs in eleven building trades.

The Home Builders Institute recruits both women and minorities for its Craft Skills Program, which is operated by more than 100 builder associations and is the gateway to the trades used by nearly 50,000 young men and women.

A number of employment programs dedicated to helping women get better jobs emerged from the women's movement. These groups offer valuable training and advice on entering nontraditional jobs. One organization, Wider Opportunities for Women, Inc., is based in Washington, DC, and is known by

Opportunities for Women and Minorities

its initials—WOW. Services include interview skills, assistance in filing applications, assessing one's skills, and exploring career options. WOW operates a program that prepares economically disadvantaged women for construction jobs.

The national network of YWCAs has a long tradition of assisting women in seeking more challenging, better-paying employment. More than a century ago YWCAs took on the task of teaching women to type—at a time when such a job was considered "too demanding" for women! Current efforts focus on identifying job opportunities, recruiting women workers, and providing necessary support services.

More than 8,500 women at work in construction comprise the Fort Worth-based National Association of Women in Construction. This organization provides job referrals, takes part in career recruitment, shares industry knowledge and ideas, and offers construction education, including scholarships for women preparing to enter the industry. It also provides resources for women who own or manage construction firms. To contact NAWIC, call 1-800-552-3506 or write to the NAWIC Executive Office, 437 South Adams Street, Fort Worth, TX 76104.

Another sponsor of job training programs is the National Urban League's Labor Education Advancement Program (LEAP), 500 East 62nd Street, New York, NY 10021.

To learn more about recruitment programs for minorities and women, get in touch with the Job Service office listed in your telephone directory.

5

Education and Training

There are a number of ways to gain the training and experience needed to launch a career in the construction industry. Some combine work and skills training; others pay wages during on-the-job training. There are also many school-based programs, for which you pay tuition and which confer a degree. High school graduation is still the basic starting place, but don't give up if you dropped out of school. The Job Corps has programs that will start you in job training while you finish diploma requirements.

Whatever path you choose requires your personal commitment to work hard and maintain a positive attitude, for both are keys to success.

Seven proven training routes are the following:

- High school vocational-technical training
- Private trade and technical schools
- Armed Forces
- On-the-job training
- Preapprenticeship and other special training
- Formal apprenticeship
- Community colleges and technical institutes

In practice, substantial overlap exists among these programs, and many of today's construction workers have taken part in several of them. For example, high school vo-tech students often acquire on-the-job training (OJT) through co-op work placements. Formal apprenticeships also include a significant amount of time devoted to supervised OJT. The formal apprenticeship, while requiring the longest commitment of time, is also a tried and true route to skilled construction jobs.

MAKING A CHOICE

As you ponder how to prepare for your career, you will most likely be concerned with limitations of money and time. Unlike many career fields, there are still ways to enter the construction field if either or both of these factors represent a problem. The discussion that follows considers each of the career routes listed above, arranged from the least costly to the most costly in time and money.

The education and experience you gain from vo-tech classes, enlisting in the military, or seeking a pre-apprentice slot or entry-level job with OJT are the "freebies." They may not fulfill your ultimate desire for job training, but participation in one or more can only increase your chances of earning an apprenticeship slot. They will also give you a chance to discover whether the construction industry is truly your vocation.

Community college and technical institutes offer associate degree and certificate programs in skilled occupations. Tuition is usually low, and financial aid is available.

Private trade and technical schools provide real-world training for people who have finished high school. Courses range from a few months to two years, focusing on a specific skill. Choose carefully, however; the tuition may be equal to or higher than a four-year college program.

Increasingly, state-related colleges and universities

have been developing four-year degree programs in construction, usually aimed at the man or woman who plans to own or manage a construction company.

INDUSTRY EFFORTS

The construction industry and labor unions associated with it have taken vigorous steps to help assure a supply of well-trained, motivated workers for the future. Such workers improve employer-employee relations; provide a pool from which future supervisors are selected; attract capable young men and women to the industry, and help to raise the general skill level of the industry.

In the residential construction industry, the Home Builders Institute of the NAHB provides four separate high-quality education and training programs.

In its Craft Skills Program, nearly 50,000 young men and women have been trained to enter the industry through preapprenticeship and apprenticeship training programs provided by more than 100 builder associations across the nation.

Community Revitalization Projects run by the Home Builders Institute (HBI) help cities train the unemployed for work in construction while rehabilitating community-owned properties. The HBI manages these training programs, with full support of local builder associations. Completed projects, often affordable housing units, are made available to low- and moderate-income families.

Correctional Training Programs are offered to offenders nearing release. By acquiring construction skills, these men and women are better prepared to reenter the work force.

THE JOB CORPS

For nearly thirty years the Job Corps has offered basic education, vocational training, and job-placement

Three participants in the federal Job Corps demonstrate their skills in replacing and installing floor coverings (courtesy International Brotherhood of Painters and Allied Trades).

services for more than two million young men and women with a common goal—to learn a skill that will give them a future in the work world. The Job Corps program got a big boost in 1974, when the NAHB joined forces in developing a program to train participants in construction trades. Its HBI trains young people for work in eleven construction trades that are in demand. More about the Job Corps experience may be learned by reading Chapter 7, Vignettes. These real-life success stories may provide just the encouragement you need to start your own career plan.

The quality of instruction offered at Job Corps

centers is unequaled. Experienced journeyworkers serve as instructors at residential camps across the U.S. In addition to receiving classroom instruction and hands-on training, participants who left high school without a diploma can prepare to take the GED as part of their Job Corps training. Many gain the necessary edge to secure an apprenticeship in highly competitive fields.

Training is offered in bricklaying and masonry, building and apartment maintenance, carpentry, electrical wiring, landscaping, painting, plumbing, and solar installation.

Most students live in a Job Corps dormitory during their training. Leisure activities, team sports, and self-improvement classes are offered. Most students receive free room and board and medical care. A small wage is paid, part of which is earmarked for savings.

Students who complete the course of study have acquired a set of professional tools and have accumulated savings to tide them over while locating that first job. In addition, experienced job-placement counselors provide supportive help for graduates making the transition, keeping in touch even after the job begins.

To find out more about Job Corps programs in construction, write to the Home Builders Institute, 15th and M Streets NW, Washington, DC 20005, or call toll-free 1-800-368-5242, extension 550. The Job Corps toll-free number is 1-800-TRAIN YOU.

Formal Apprenticeships

The construction industry is a leader in apprenticeship training. In fact, more than half of all apprentices in registered programs are preparing for jobs in construction. This long-established tradition has proved its effectiveness at channeling well-trained workers onto the payrolls of companies who need them.

The duration of apprenticeships varies significantly from one trade to another, but programs range from one to six years, with three to four years being typical. As you might expect, there is a great deal of emphasis on job safety.

During training, apprentices start at about half of the skilled worker's starting wage, with increases of about 10 percent every six months, until by the end of training they receive about 90 percent of journeyworker pay.

As you might imagine, there is considerable competition for free job training that pays a wage while you advance toward a well-paid career. Candidates for apprenticeship slots may have to take a test and participate in a personal interview as part of the selection process. Most also require a physical exam.

Most apprenticeship applicants take the Specific Aptitude Test Battery, which has nine parts. Typically, two test areas are administered, chosen from the following list: general learning ability, verbal aptitude, numerical aptitude, spatial aptitude, form perception (the ability to perceive small details), clerical perception (the ability to distinguish important details), motor coordination, finger dexterity, and manual dexterity.

The aptitude test helps define a candidate's specific skills and abilities, an effective method of matching people with job training they are likely to enjoy and in which their chances for success are high.

Since employers and unions put up the money to operate apprenticeship programs, and employers divert the time and talents of skilled workers to train apprentices, they are interested in recruiting the best and the brightest candidates.

Apprenticeship programs are also subject to U.S. Department of Labor regulations concerning equal opportunity. Women, blacks, Hispanics, and other minorities must get a fair share of apprenticeship slots.

Some trade school graduates have an advantage in applying for apprenticeships, since they are already familiar with shop safety, materials, and tools. They may receive higher starting wages and face a shorter training period.

A solid foundation in reading, writing, and math is required of all apprentices. Candidates who have done well in high school shop math, drafting, and physics also have an edge.

In addition to education and physical fitness, apprenticeship candidates are evaluated on the basis of interest shown, attitude (toward hard work, authority, and teamwork), and such personal traits as appearance, assertiveness, sincerity, dependability, character, and habits.

Points are assigned in each ratings category, totaling 100. Scores for the interest, attitude, and personal traits sections come from the personal interview. The total point scale is entered on "the register"—a waiting list for apprenticeship openings, and the candidate is notified of the scores given.

The next step is to wait for an opening. Job experts say the wait can be a few months or even years, depending on the number of openings and the number of applicants with ratings higher than your own. Estimates indicate there are eight applicants competing for every apprenticeship slot.

The state of the economy and geographic location also have an impact on apprenticeship availability. As you might imagine, when skilled workers are being laid off, the chances for trainees to get a start decline sharply.

If you are not called within two years, you have to reactivate your file by reapplying. That does not mean you have to go through the interview again, but chances are that you'll improve your score if you do.

EDUCATION AND TRAINING

THE APPRENTICESHIP AGREEMENT

Before OJT or classroom work begins, a formal apprenticeship agreement is signed by the trainee and the trainer or training organization. Apprentices who are under 21 years of age must have a parent or guardian sign the agreement. This document legally assures the apprentice of training for a career as long as the apprentice lives up to its terms. You receive an ID card, which is updated regularly to reflect your progress toward journeyworker status and wages. The contractor who employs you puts you on the payroll. An effort is made to have apprentices receive fringe benefits on the same basis as journeyworkers.

If some circumstance forces you to drop out of the program, you can resume training later with credit for the work already done.

According to the U.S. Department of Labor, 37,033 apprentice electricians were learning the craft in 1990. Of that total, 35,387 were male; 4,951 were members of minority groups, and 1,646 were females. In the ranks of 27,206 apprentice carpenters, 25,795 were males, 4,608 were minorities, and 1,411 females.

TYPICAL ELECTRICIAN'S APPRENTICE TRAINING

Those preparing to become electricians must understand the nature of electricity: how it is generated, transmitted, and controlled. In the past some electricians became qualified by working as helpers. This OJT and time spent studying on their own was sufficient preparation for passing the journeyworker's licensing exam. Today's highly specialized work requires specialized training under an approved program.

Apprentices are required to master a wide variety of mathematical formulas and computations. Becoming an electrician can be interesting and challenging for those who enjoy physical as well as mental achievement.

Candidates for apprenticeship are evaluated on the basis of academic performance and potential, health status, attitude, and enthusiasm for the work. During the course of their training, apprentices usually work for several contractors, becoming familiar with a variety of electrical jobs.

The Construction Electrician's apprentice program covers four years (8,000 hours) and is a combination of on-the-job training and classroom instruction, the latter covering 144 hours per year. You will take courses in electrical theory, the basic principles and measurements of electricity, AC and DC current, and basic electrical devices and circuits. You will learn to read blueprints and study the types and sizes of wires, connections and joints, different methods of wiring electrical systems, service entrance and branch circuits, and installation of switches and other devices. There will be study related to planning and estimating a job, and many programs offer some study of electronics. Fundamentals of the electrical code, which promotes safety in electrical installations, are also covered.

While on the job you will be guided by an experienced electrician. At first the tasks will be routine—learning to drill holes, set up conduit, and set anchors—while you learn to handle the tools of the trade. You will progress to measuring, bending and installing conduit, and connecting and testing complete wiring systems.

The first 2,000 hours are devoted to Residential Electrical Systems, including single-phase service and metering, remodeling, installation of equipment and appliances, and installing light fixtures, receptacles and switches, and security systems.

Phase II covers commercial-industrial applications and encompasses 5,000 hours. Apprentices learn about both 240- and 480-volt electrical systems. They study metering, polyphase and current transformers, installing

Education and Training

conduits and outlets, installing wiring beneath concrete slabs and masonry, steel construction, buss-duct systems, under-floor ducts and metal raceways, and installations that are vaporproof and explosion proof. Other topics include circuiting for light and power, various types of motors and their controls, and transformer applications and connections.

Phase III, devoted to specialized work, involves 1,000 hours. Apprentices learn about welding, employee-management relations, customer and employee relations, and electronic, communications, and fire-alarm systems.

When an apprentice has successfully completed this work, he or she is eligible to sit for any required licensing exam and may also apply for membership in the International Brotherhood of Electrical Workers, which represents more than one million members across the U.S. and Canada.

Typical Carpenter's Apprentice Training

By the time a carpenter's apprentice achieves journeyworker status, he or she will have invested nearly 8,000 hours in learning the trade. Both at the job site and in the classroom, the instructor will be a professional.

Here's how the typical apprentice spends that time:

Shop training: 1,500 hours (approximate), including use of hand tools and woodworking machines, layout, and planning. Construction and installation: 2,300 hours, including buildings, machine footings, forms, production equipment, office construction. Maintenance and repair: 2,028 hours, including buildings, production equipment, and office. Advanced carpentry; 700 hours, including building furniture; 800 hours of optional training, plus 80 hours of safety training. Classroom instruction includes 180 hours of math, 72 hours

Education and Training

of science, 108 hours of shop instruction, 144 hours of drawing, and 72 hours of electives.

Those 8,000 hours represent four years in the life of a construction apprentice. Regardless of the specialty, the time is divided into eight six-month periods of continuous employment (so far as economics and weather permit), during which a probationary period is worked first.

Apprentices with documented training and experience may be granted advanced standing after demonstrating their skills.

The National Joint Carpentry Apprenticeship and Training Committee sets standards for all recognized apprenticeship training programs. This committee reflects input from the United Brotherhood of Carpenters and Joiners of America, the primary carpentry union, plus the Associated General Contractors of America and the National Association of Home Builders. The U.S. Department of Labor also oversees apprenticeship training.

Performance-Evaluated Training

More than a decade ago the Carpenters' Union pioneered an audiovideo training method that has proved effective in apprenticeship. The basic components are likely to be found in other types of construction skills training. Color slides are used to demonstrate step-by-step the task being described. The apprentice then has an opportunity to perform the same task under the watchful eye of an experienced supervisor. Tasks are divided into skill blocks, each covering the necessary

Workplace safety is a big part of apprentice training. Here carpentry apprentices learn to work overhead on a stairwell (courtesy Home Builders Institute).

blueprint reading, safety procedures, and use of tools. The apprentice who completes the required number of skill blocks (for carpenters, six) is advanced in educational position and usually receives a pay raise.

COMMUNITY COLLEGE OR TECH SCHOOL

A number of public and private schools offer high school graduates specialized training in construction trades. Typically, such programs combine classwork, fieldwork, and safety instruction. Graduates typically receive a certificate of completion. Community colleges award an associate degree if the program includes 50 to 60 credit hours of study.

Community colleges are usually the most affordable. Students undertake supervised building projects on campus or take part in cooperative education including hands-on experience with a contractor as part of the course. If you are considering enrolling in a private technical school, ask local contractors whether the school meets their requirements for entry-level training. Also, the tuition may be substantially higher than that of a community college.

Associate Degree Programs

Let's take a closer look at associate degree programs in building construction technology for the would-be contractor and in the skilled trades of construction carpenter, electrician, plumber, and landscaper. With work experience, each of these trades could lead to a position as a supervisor or form the basis for launching a business of your own.

Since the ability to communicate is essential in all building trades, students must demonstrate the ability to write letters of application, memos, work orders, and reports and use communication skills on the job.

Education and Training

Building Construction Technology. This degree program focuses on residential and commercial construction, emphasizing job layout, time and cost estimates, and project management. Courses in English, math, science, computer applications, and economics are included to enhance a student's career opportunities. The credits are transferable should the student decide to earn a bachelor's degree.

With such training and suitable job experience, a worker would qualify for jobs leading to supervisor, contractor, construction technician, or construction superintendent.

The high school program recommended as a minimal prerequisite includes two years of algebra, one of geometry, and one of science. One year of advanced algebra is desirable.

A graduate of building construction technology should be able to:

- Write clear, concise, legible, and accurate technical reports and use verbal communication skills in job-related activities.
- Demonstrate basic manipulative skills needed to lay out and plan work.
- Interpret plans, drawings, specifications, lines, symbols, and abbreviations on working drawings or blueprints.
- Demonstrate the ability to lay out and erect residential and commercial structures.
- Analyze specifications and contract drawings; make accurate quantity takeoffs and labor estimations to develop an estimated cost for constructing a project.
- Prepare preliminary architectural drawings and sketches.

- Describe various types of materials and methods used in the construction trade.
- Demonstrate basic knowledge and skills in masonry and concrete construction.
- Describe the organization, financing, labor relations, selling, pricing, customer service, management, and other aspects of business.
- Describe the relationships among the various trades.
- Solve building construction problems; problems using higher math; apply scientific procedures to construction problems; apply technical and basic skills to practical construction; apply the microcomputer to construction applications; practice safe work habits; demonstrate responsible attitudes and high-quality work.

Construction Carpentry, a course devoted to both rough and finish carpentry. In rough carpentry students learn to erect frameworks for buildings, subfloors, and rafters and build forms. Finish carpenters learn to install molding, cabinets, and door and window trim and to build stairs and floors.

Both phases of instruction emphasize the correct use of tools including hand tools, portable power tools, and power-activated tools. Job layout and design, correct measurement and selection of materials, coordination with other workers, and construction techniques and technology are other topics explored through class activities and supervised construction projects.

Students may specialize in either carpentry or home remodeling midway through the course. The carpentry option includes instruction in concrete and commercial construction; remodeling includes home remodeling, plumbing, and electricity for the trades.

Electrical Occupations. The course offers skills and

EDUCATION AND TRAINING

theoretical background needed for a variety of careers. Graduates may work as electricians in electrical construction or in electrical maintenance. The graduate is prepared for a job as a construction union apprentice, an electrical inspector, or self-employment in residential and commercial wiring.

Students work in a construction lab, learning about direct current and alternating current, basic electronics, engineering drafting, accident prevention, electrical machinery analysis, troubleshooting, and testing.

They learn to demonstrate technical skills in a variety of electrical fields, apply safety standards, meet work quality standards, and demonstrate knowledge in electrical theory, math, and physics as they apply to construction.

Students also learn the use and care of electrical tools and materials. They learn to read and develop blueprints to perform installations that comply with the National Electrical Code. They are expected to: demonstrate the use of troubleshooting equipment and standard testing procedures; set up ladder relay logic systems and convert them to electronic programmable control systems; and demonstrate knowledge of basic electronic control circuitry, devices, and schematic diagrams. They also learn to interpret ideas and develop plans through communicating with others.

Plumbing. This is generally a one-year course that covers basic theories of plumbing, soil waste and vent layout, household and industrial maintenance, sewage systems, and the use of hand and power tools.

Students develop skills in all types of plumbing repair work used in residential, institutional, and commercial applications. The course also provides training in fundamentals of communication and math and prepares students for jobs in plumbing installation, industrial maintenance, public utilities service, and shipbuilding.

Students also receive an introductory course in refrigeration and heating, air conditioning, and ventilation. The basics of electric theory, interpreting blueprints and building codes, carpentry for the trades, and gas and electric welding are also covered. Students learn to lay out, estimate, calculate, and use math skills required in the trade; install, maintain, and repair plumbing systems; identify the principles involved in the collection, storage, and use of solar energy for space and domestic water heating; and apply conservation measures to plumbing installations.

Landscape Construction is a program focusing on techniques used to build landscape features. It includes construction of patios, walks, retaining walls, fences, fountains, waterfalls, pools, and steps. Students learn about job specifications, bidding and pricing landscaping jobs, basic surveying techniques, drainage, and grading.

In addition, coursework covers landscape and ornamental plants, design applications, and development of planting and maintenance plans; and the study of soils and fertilizers, synthetic soils, and techniques used to control insects, disease, and weeds. There are also courses in landscape planning for both private and public land use.

Bachelor's and Master's Degree Programs

Colleges and universities offer undergraduate and graduate degree programs to prepare people to own or manage a construction company. Students are likely to be the sons and daughters of families already involved in construction, or adults with experience in the field who want to move up to management or launch their own business.

Core programs concentrate on construction materials and a broad study of building methods, structures,

Painting the exterior of this warehouse is all in a day's work for Job Corps Program participants (courtesy IBPAT).

environmental systems, and management techniques. Graduate degrees in building construction or construction management have also begun to appear.

The NAHB has launched a student program with chapters on nearly 100 college campuses where construction degree programs have been established. A list of NAHB-affiliated schools appears in the Appendix, as well as colleges and universities granting degrees in building and construction technologies.

6

Entering the World of Work

How will you know whether a construction trade is the right career for you? It has many attractions, not the least of which is work that pays well and offers good benefits. Still, there is always a downside. Before committing to training, it would be wise to spend some time thinking about the pros and cons. If in your mind the advantages far outweigh the disadvantages, you are much more likely to be satisfied with your career choice over the long haul.

First, consider what attracts you to the construction trades. Perhaps it is the physical challenge, the sense of accomplishment, or that you like being outside and for the most part doing your own thing.

No matter how much care and attention you give to learning about a career in advance, professional job counselors agree that you cannot be one hundred percent certain that you will like a job. Sometimes actual work experience is the only way to learn for sure. Still, you can improve the odds by asking yourself some realistic questions.

For example, how is your overall health? As a group, blue-collar workers are healthier than the general population, and it's a good thing, considering the strenuous tasks they are often asked to perform.

Would you be willing to take a complete physical

Apprentices often perform community service projects during their training. Here a team of apprentice painters help preserve a historic covered bridge (courtesy IBPAT).

exam? Most hirings depend on the results of such an exam. If you are hesitant, it will reduce your chances of being hired.

Do you have reliable transportation? Many job sites are on the outskirts of town or perhaps in a town 50 to 100 miles away.

Can you budget money for a rainy day? That is important in construction work, since at least in the early years you'll be the first laid off if work becomes scarce. Construction workers earn pay only when they're on the job. Rain and material delays can shut down a job site. There are also downtimes between projects and during the off-season. Unemployment and sub-pay certainly help, but they do not totally replace wages.

What type of work atmosphere do you find most comfortable? As in many fields of work, newcomers are often subjected to pranks from the more experienced workers. While you are low in seniority, you will probably be given the lousiest jobs; with seniority, you will be able to bid jobs that pay better and are usually easier to perform.

Are you ready to face rugged working conditions? Remember that you may work outdoors in all sorts of weather. Climbing ladders, working from a scaffold, carrying materials or tools—these are all part of the game.

Two personal qualities that will mark you for success early in your career are a high level of self-confidence and the ability to stick with a project.

APPLYING FOR A JOB

A union journeyworker's card is the surest guarantee of work in the construction trades. No one is issued such a card unless he or she has demonstrated skill and gained considerable experience. Aside from union hirings, the construction industry relies on two standard employment tools, the job application and the personal interview. The résumé, a formal typed history of your educational and employment background, is not a requirement for job-hunting in this field. Still, you may find it easier to complete the job application by referring to such a chronological list.

It is a good idea to read the entire application before making any marks on it. Crossed-out lines, mistakes, and erasures reflect negatively on you. Complete each item fully and truthfully, assuming that the information will be verified. If you have any questions, ask for clarification before you attempt to answer. Sign the form if requested to do so. Also, check the back of the form; it may contain questions as well.

Entering the World of Work

The Interview

Most hiring decisions rely heavily on the results of a job interview. That involves one or more face-to-face meetings between the candidate (you) and someone who represents the employer. It is natural to feel a little nervous before a job interview, but there are ways to prepare yourself for the experience, thereby boosting your self-confidence and helping you make the best impression on the interviewer.

Start by knowing about the job you want and the company in general. If you know someone who works for the company, pick their brain for inside information. Before your interview date, stop by the office to obtain copies of company brochures or annual reports, or visit job sites to see what kind of structures they build. Why? You will be expected to ask some questions during the interview too. These questions should show whether you want the job enough to take an interest in the company before you are hired. Your knowledge also reflects initiative, another positive trait. If you sit in silence or fumble for an answer when the interviewer asks a question, you will probably lose points.

Yes, asking those questions is difficult. You should practice them in advance so that you sound relaxed and in control of your thoughts during the interview. What topics are appropriate? You can't go wrong if you ask about:

- The job duties, working conditions, and opportunities for advancement.
- Qualifications they are looking for: physical, emotional, and educational.
- Fringe benefits.
- Working hours.
- Will you work alone, or in a group? How is the group organized? Whom would you report to?

- When will you know if you've been hired? What are your next steps?

Ask the questions that most interest you. These are simply suggestions. You may want to add specific questions.

Another way to feel more comfortable is by preparing for the interviewer's questions. It is possible to anticipate them. For instance, many interviewers will ask:

- Why do you want to work for our company?
- Why do you think you are qualified for this job?
- Tell me about yourself and the work you like to do.

Later in the interview you might be asked about your family, how you spend your spare time, what you think you might be doing in five years; what you consider to be your major strengths and weaknesses; the subjects you liked best in school.

Put your thoughts on paper, and while you're at it, consider a few more commonly asked interview questions: What did you do on your last job? What makes you feel you are ready for this job? What are your best successes and accomplishments? What made you leave your last job? How much money do you feel you should be earning? Why should we hire you instead of someone else? How long do you intend to stay with us? What would you like to tell me that is not on your job application?

Practice a response for each question until you feel comfortable with it. Be sure your response is genuine and shows something of who you really are. All the while, remember to emphasize your strengths and don't forget to admit, yet downplay, your weaknesses.

The Big Day

To give yourself an edge, be neat, clean, and alert. Plan to get a good night's sleep, and set the alarm a little early. Wear appropriate clothes. For a construction job, choose clean, presentable jeans and a dress shirt. A recent haircut is advised, and don't smoke in the reception room or during the interview, even if invited to do so. If you have a completed application, references, and so on, make sure your name, address, and phone number appear on every page. Place the unfolded papers in a large envelope or clip them securely in a folder.

Don't be late; in fact, try to arrive five to ten minutes early. Tell the receptionist your name and whom your appointment is with, then take a seat and try to appear calm. Don't shuffle through your paperwork or do anything unusual; the receptionist may have been asked to observe the behavior of job candidates.

Remember to look the interviewer in the eye, which is not to say that you should stare. Lean forward a little, and demonstrate that you are a careful listener. Don't fidget. An occasional gesture is fine; otherwise keep your hands in your lap or on the arms of your chair. Your voice should be clear and assertive in tone. As with gestures, be conservative with your smile, but don't go through the whole interview with a poker face. Above all, don't fall into the trap (sometimes deliberately set) of complaining about past employers.

In today's economy, only the largest corporations can afford full-time professional interviewers. Remember that the person you talk with may be as uncomfortable in the interview situation as you are.

You may be asked to take one or more tests before the interview begins. Construction workers are frequently given tests that measure mechanical aptitudes or math skills. Remember that people who are tired or don't feel well generally do poorly on tests, but sometimes being a little nervous actually helps focus your attention. Just read or listen carefully and follow directions.

In today's competitive market many job-seekers face a few turndowns before scoring a success. By taking time to prepare for your interview, you will certainly improve your chances of being hired. Remember that today's jobs demand workers with initiative, people who can solve problems and follow directions. Your preinterview work will sharpen skills in each of these vital areas.

Most interviews end with a noncommittal comment such as "Thanks for coming in; we'll get back to you." You may leave with no idea how you did or whether you have a shot at the job.

Should you simply wait for a phone call that may never come? Recruitment counselors recommend that you send the interviewer a short letter within a day or two, thanking the person for seeing you and mentioning again your interest in the job.

A week or two later it is permissible to make a follow-up call asking about the status of the job. Has it been filled? Are you still being considered? If you call a few times and the response is less than encouraging, it is time to apply elsewhere. Even if you don't get the first job you go after, having been though the interview process is a valuable experience that will help you on the next go-round.

7

Opportunities for Advancement

Many craftsworkers receive enough challenge as a journeyworker to last a lifetime, yet for some there is always a challenge to be met and a new goal to strive for. If you are the ambitious type, you are probably already wondering what opportunities for advancement lie ahead. A skilled craftsworker with several years of experience may aspire to one of the advanced or supervisory positions discussed in this chapter. In general, earnings reflect the experience the worker brings to the task and the increased responsibility. Although wage rates vary widely across the country, it is fair to calculate a pay scale 10 to 25 percent higher than the journeyworker rate. When the employer is a big business, many of these jobs include such "perks" as the right to drive a company vehicle to and from work, and participation in a profit-sharing plan.

As you plan each step of your future, remember that ambition, drive, and a willingness to adapt to change lie behind successful career moves.

SUPERVISORY POSITIONS

Crew Chief

Those who seek advancement would be likely to set their sights first on the job of crew chief or crew leader, a position called "foreman" in the days before gender-neutral vocabularies. This man or woman directs the day's work, clarifying what is to be accomplished, and assigning duties if need be. The crew chief is always aware of the schedule to be met and serves as a troubleshooter, keeping the work at a steady pace. Many crew chiefs take an active role in the work, and all report to the job superintendent or the contractor.

A crew chief should be a "people person" with excellent communication skills, an understanding of human nature, and the ability to motivate workers to their best performance.

Job Superintendent
On small to medium-size projects, a job superintendent is often named to direct all construction activities or specific phases of major projects. Direction of work crews is done by communicating with the crew leaders.

General Superintendent
This supervisory person receives directions from the project manager, the job superintendent, and often the subcontractors. The job depends heavily on communication skills and the ability to follow up, making sure that tasks are accomplished on time and meet performance or quality assurance standards.

Project Manager
Representing another step up in responsibility, this person directs all construction functions for a large project, usually starting with plans and specifications. The project manager sets the work schedule and establishes procedures and job policies.

Estimator
With this job, we take a step away from construction site management. Estimators operate partly in the world

of finance and partly in the construction business. To do their work, they employ an understanding of the crucial interrelationship of time, labor, and money. The work is both interesting and challenging, and every company engaged in contract work has at least one person who functions as job estimator. The task may fall to an experienced supervisor or a specialist, although in small companies it is often the owner who fills the role.

Medium-sized to large companies generally find it more efficient and profitable to have all contract costs estimated by someone who does that work only. Good estimating skills require a blend of business knowledge and practical experience, plus skill in using the tools of an estimator's trade—tables, charts, and graphs. With them, the estimator translates blueprints and job specifications into material orders, a labor force of the right size and skills, and a reasonable work schedule.

Most job estimators divide their time between desk duties in the office, site visits, and meetings with architects, engineers, and the like. During the building season an estimator works at a demanding pace and at such times may encounter work-related stress. Because controlling costs is such a vital part of the estimator's work, many receive profit-sharing as an incentive.

Expediter
This person keeps the job on schedule by reviewing deliveries, scheduling material deliveries and work crews in a timely fashion, obtaining necessary permits and clearances, and establishing priorities on the job.

Civil Engineer
A number of surveyors and survey technicians have climbed the educational ladder to become civil engineers, the professionals who design highways, public works projects, and large construction jobs. The coursework

required to become a state-licensed land surveyor provides a foundation from which an engineering education may be launched. Others in the construction trades have great potential for engineering too, but it will probably take them longer to acquire the academic background in math and science. Depending on the source of employment (private industry or government) and region of the country, starting salaries for engineering professionals are in the neighborhood of $27,000 and may increase to near $50,000 with experience.

Inspectors

Because quality control is such a crucial aspect of construction work, various inspectors are charged with the responsibility of examining the project as the work progresses. Armed with specifications and checklists, they must confirm that the work complies with building codes and ordinances, zoning regulations, and specifications of the contract.

A background as a journeyworker and experience in a trade can lead to assignment as an inspector. The field has various specialties including building, electrical, elevator, mechanical, plumbing, public works, and home inspectors.

The first inspections are made during construction, with periodic visits to see that regulations are being observed. Construction and building inspectors may use a computer to aid in record-keeping, storing the thousands of details required to monitor a large construction project.

The category of **building inspector** includes additional specialization. Some concentrate on structural steel, others on buildings of reinforced concrete. Before building starts, they must inspect soil conditions and the position and dimensions of footers. After the foundation is completed, another inspection is made.

Opportunities for Advancement

The number of further inspections depends on the size and type of structure. Once completed, a final inspection is made, which is usually quite comprehensive in scope.

Electrical inspectors are concerned with the proper installation of electrical systems and equipment, checking for proper function and compliance with electrical codes. They look at new and existing wiring, lighting, sound and security systems, generating equipment, and any electric motors used in the building. Electrical wiring used in installing air conditioning, appliances, and similar systems must also be inspected.

An **elevator inspector** must assure that all lifting and conveying devices are safely installed. The title of this job is somewhat deceiving, since in addition to elevators it also covers escalators, moving sidewalks, ski lifts, and various amusement rides.

Mechanical inspectors deal with such devices as commercial kitchen appliances, oil- and gas-fired appliances, boilers, ventilating equipment, heating and air-conditioning units, and the piping systems that carry natural gas or oil.

Plumbing inspectors check that all codes and requirements have been met in the installation of water supply and waste disposal systems. They examine plumbing fixtures, traps, and drains along with waste and vent lines.

Public works inspectors are concerned primarily with large government-sponsored projects such as highways, bridges, and dams. They check to see that contract specifications have been met and keep records of the work done and materials used to aid in calculating contract payments.

Home inspectors are often employed by municipal governments and inspect newly constructed houses to assure that they comply with building codes, zoning requirements, and the like. Sometimes these inspectors are hired by real estate brokers or prospective homebuyers. In such situations the structure may not be new. After the inspection, a detailed report is written to document the condition of all major systems.

All building and construction inspectors rely on what they see and what they know about construction standards to perform their duties. Often they use test equipment, meters, tape measures, and surveying tools to compile their records. A daily photographic log is sometimes employed.

Should the inspector discover a detail that does not conform to the building code or project specifications, the contractor or another person in authority at the job site must be notified immediately. If corrective action is not taken within a reasonable time, government inspectors may call a halt to all work on the project.

Inspectors work alone on small to medium-sized projects but may function as part of a team on large jobs. Their time is divided between inspection visits to the work site and office duties such as telephone conferences, reviewing blueprints, and writing reports. Inspectors usually work a 40-hour week.

The salary range is wide, with the lower-paying jobs found among inspectors who work for local government agencies. Experienced construction and utility inspectors may earn from $40,000 to $70,000 per year, plus benefits. A company vehicle is commonly supplied.

8

Details to Consider

Like any other career choice, these jobs have many good points and some disadvantages as well. To make a good career decision, you need to consider them carefully.

More than 40 percent of all construction trades workers are younger than 30. Studies show that they have a tendency to move into other fields once they reach their thirties. What jobs do they move to, and why? The answers to those questions may have a bearing on your own career direction in ten or fifteen years. Still, the fact that there is considerable job turnover in the industry means that plenty of job opportunities are available for newcomers.

One undeniable fact about construction work is its cyclical and seasonal nature. As workers assume increased economic responsibilities, they may be unwilling to deal with seasonal layoffs. The effect of bad weather conditions is greater in construction employment than in other types of industry. This story is not totally one of gloom and doom, however. In many snowy regions new building methods and temporary weather shelters extend the building season.

The weather is not so much a problem in the South and the West, but not all workers are willing to relocate, and the trade-off of working in excessive heat is something to consider. Even in areas where construction goes on year-round, there can be downtime between jobs.

You must also be aware that construction work generally requires great physical stamina and strength. As workers age, they may become less able or willing to tolerate the harsh working conditions found on construction jobs. The combination of these factors may contribute to older workers' decisions to leave the field. Still, many of them use the skills they have acquired to develop a new career, often construction-related.

We have already mentioned the dirt, noise, and dust typical of work sites. Modern protective devices dampen construction noise yet permit you to hear what you need to for safety's sake. Speaking of safety, you may already have gathered that a construction site can be hazardous to your health. Steel-toed shoes, a hard hat, and safety goggles or a visor are standard issue. Many jobs require being comfortable working a considerable height off the ground, using scaffolding or staging as a platform for you and your tools. Workers fall, and sometimes things fall on them.

Construction workers often lift and carry heavy tools and materials. A significant number of older workers report chronic back problems. These people also work around sharp tools, hot pipes, soldering equipment, electricity, and natural gas lines. Some jobs require standing on hard surfaces for a long time or working in cramped places. There are days when you work under a hot summer sun, or in cold rain or sleet. The building season comes to rule your lifestyle. You may never again take a summer vacation, for those weeks are the height of the season. Independent (nonunion) workers rarely receive paid vacations, and paid sick leave is also rare. You may work overtime for weeks at a stretch, and then be laid off, waiting for the next job to start.

Many workers, operating engineers in particular,

For the story of how one woman established her own business as a residential painter, see Chapter 10 (courtesy IBPAT).

must consider the constant vibration of running a machine over rough ground. Still, that is more of a problem with older machines than with the newer models. You may be fortunate enough to run a machine from the relative comfort of an enclosed, climate-controlled cab.

Throughout the workday you must rely on a combination of talent, experience, and problem-solving skills to meet the demands of your job.

On the other hand, construction wages may be double or more those of comparably trained workers in shops and factories. During layoffs the prudent construction worker has savings, unemployment benefits, and often sub-pay from a union to make ends meet. To keep their best workers, some contractors guarantee a certain number of weeks of work.

Although construction jobs are to be had in most communities, if you are in a union or work for a large company you may be assigned to a project many miles from home. During those weeks or months you live in motels through the workweek, perhaps with a coworker as roommate to cut down on expenses. Some projects last six months or more, and you may decide to rent a house or apartment and move your family nearer the job site. Of course, if you have children of school age, this nomadic life may not be practical. If your family stays behind, the costs of maintaining separate residences, long-distance phone calls, and weekend trips home all come out of your paycheck.

Still, imagine the satisfaction you will feel the first time you drive your family down a brand-new superhighway that you helped to build. The challenge and reward of producing a tangible product through your own efforts is becoming a rare commodity as America shifts into a service-based economy. Bridges, roads, high-rise buildings, dams, and airports last a long time and stand as monuments to the skilled craftsworkers who built them.

Obviously, every job has plus and minus factors. It is a good bet that few people are 100 percent satisfied with their work at all times. The trick is knowing in advance what tradeoffs you will be asked to make, then deciding how much the career means to you.

Personal Qualifications

The construction careers we have looked at are unique in some ways, but they also have certain features in common. The work is performed outdoors, and it requires people with mechanical ability, manual dexterity, and the ability to follow written or oral instructions.

Details to Consider

Construction workers are men and women with stamina and the strength to do active, often strenuous, work. Good health, keen eyesight, physical coordination, and alertness—all these are certainly assets. Several of the jobs, particularly in structural iron and steelworking, demand a good sense of balance and the ability to judge distances. Success also depends on your respect for safety factors and ability to take care of equipment placed in your care. Much coordination of schedules and teamwork goes on behind the scenes at a construction site, so it is important that you get along with fellow workers and supervisors.

Think about the work atmosphere in which you are most productive and comfortable. This is a male-dominated field; although the barriers are coming down, women in nontraditional careers still must earn the respect of coworkers. Don't sacrifice your dignity. Let the quality of your work and a positive attitude speak for you.

How good are you at sticking with a project when the going gets tough? How confident are you in your own abilities? The natural process of maturing will help you develop in both of these areas, but by being consciously aware of your abilities and striving to improve them, you will make even bigger gains.

If You are Still in School

What courses should you be taking in school? Math (especially algebra), earth science, computer literacy, drafting, and shop courses that teach you to use and care for tools are all useful preparation. If you have access to vocational-technical courses, you might consider blueprint reading, construction technology, masonry, or warehousing, where you could gain practical skills. Don't forget English, speech, or communication classes, since all construction jobs require that you understand and follow directions, written or oral, and work well in a group.

There are several physical and emotional requirements you could work on now. Stamina, physical fitness, and a certain degree of upper-body strength are required in many of these jobs where the pace is fast and the work strenuous. You may be required to work long hours in the hot sun.

In many blue-collar occupations rookies are subjected to a certain amount of teasing from the established workers. The young person who demonstrates a willingness to work and an even temperament will get through this phase with the least irritation, earning the respect of fellow workers.

How is your track record for school attendance? Believe it or not, it says a lot about the attitude you will display toward your job. Employers, the people who award apprenticeship slots, and recruiters for the military and higher education all view a good school attendance record as a strong indication of commitment. Absent workers cost employers money. Likewise, training programs pack a lot of material into a day. In the interests of producing valuable employees, these slots go to the people most likely to show up and learn.

So there you have it. If you're in school, attend regularly and pay attention, especially in those classes that lay the groundwork for your future. When you're not in school, go after work experience that will give you a clear look at the career you seek while putting you ahead of the competition for job-training programs or an entry-level position.

Getting Started

If you think you have discovered a career you would enjoy, don't wait another day to start planning for the future. If you're in school, stay there and commit

Details to Consider

yourself to hard work, especially in the courses listed earlier. Acquire practical work experience, all the while building a reputation as a reliable, hard-working person who would benefit from training. Decide on a long-term goal, then set several intermediate goals and decide what steps you must take to achieve them. School guidance counselors and job counselors at your local Job Service office can help you.

Don't let current shortcomings stop you, for they can be overcome, regardless of your age and how much money or education you have. Job security, a sense of accomplishment, and skills that will last a lifetime are all within your grasp. Take the first step; the world awaits you.

9

Becoming Your Own Boss

Many Americans dream of becoming their own boss, and that goal is quite possible to achieve through the building trades, especially if you are interested in housing or light commercial work. Still, the transition from employee to self-employment or employer should not be taken lightly. Careful preparation is required.

First you must be confident that you know all about the business you intend to enter. Then you should learn the basics of management, for as your business grows you will have both employees and an office to supervise.

As in any new business venture, in the early stages you may have to rely at least partly on savings to meet income needs. Even if you make money right from the start, you may need funds to buy materials and equipment, pay for legal advice or advertising, and so on. Good financial planning will protect both your business and personal credit ratings, another essential of a successful business.

Because so many construction craftsworkers start businesses related to housing or commercial construction, this chapter focuses on those aspects of contracting.

THE FIRST STEP

To become a contractor, a journeyworker must be prepared to bid competitively with other businesses, and to face the possibility of financial losses caused by the seasonal nature of the work. Material shortages may also complicate meeting deadlines. In all these situations experience in estimating time, material, and labor requirements helps to protect a new business venture during its infancy.

You may set your sights on constructing entire housing developments, single-family dwellings, or—a growing trend—developing a specialty service such as drywall installation, roofing, electrical services, or building and installing custom kitchens and baths.

At least in the early days, you will have to be both businessperson and merchant. At the core of all construction are the people who supervise putting together many small parts to create a finished building. You may know them as builders, contractors, and manufacturers and their dealers. Your entry point as founder of your own business is likely to be through one of these construction or construction-related activities.

Opportunities for experienced workers to advance into owning their own business are great in the construction industry because construction requires a relatively low investment of capital, operating space, and full-time personnel.

The move takes many forms. It is a logical step to go from construction superintendent to contractor. Others start with a clerical job in a contractor's office to learn estimating and purchasing. An increasing segment of business owners start with college training in architecture, engineering, building construction, and business administration.

The business a construction worker opens is generally as builder or as contractor. Contracting has various specialties, usually reflecting the owner's original craft training.

A builder is responsible for functions such as purchasing and developing land, design, and acquiring construction money and final mortgage loans, in addition to constructing the building.

Building contractors generally specialize in residential, commercial, or remodeling work. Luckily for entrepreneurs, the nature of the U.S. housing market favors the small builder. No national market exists for homes as it does for car manufacturers or the clothing industry. Housing markets are defined locally and are scattered across thousands of communities. Geography, climate, and tastes in architecture influence the sort of housing buyers seek. Time and distance also play a vital role in determining the market area a builder can reasonably serve. The farther the building site is from a contractor's headquarters, the more costly it becomes to transport workers and materials. Such ventures are generally limited to the heavy construction industry, where the additional costs are built into the bidding process.

One factor that definitely works in the small builder's favor is the ability to control overhead costs. A remarkable number of these companies are one-person operations, where the owner wears all the hats. Even after the venture begins to grow, the owner is likely to be working as the job supervisor and may even wield a hammer, string a wire, or bend a pipe, depending on his or her original trade.

A word we hear a lot these days is entrepreneur, borrowed from the French and meaning someone who takes the risk of organizing and running a business in hope of making a profit. In the beginning, an entrepreneur works long hours, at the job site by day and meeting with clients or handling paperwork in the evening. Designs and building plans are kept in tune with the needs of the builder's customers, because he or she lives in the community and has a good perception of what buyers want.

TYPES OF HOME BUILDERS

Home builders operate in three broad categories: speculative builder, custom builder, and upkeep and improvement builder. The state of the economy and the financial circumstances of the business determine which category an entrepreneur is in at different stages in running the company.

The speculative builder assumes all financial risk for the house or project, generally buying the land, then borrowing on its worth to secure materials. The risk remains until the house is sold. The longer that takes, the more the profit is eaten away by interest on the construction loan. Except in severe economic downturns, the "spec" builder can survive and even flourish.

A custom builder faces less risk, but the pool from which to draw customers is generally smaller. In this operation, individuals or companies ask the builder to submit a bid to construct a house or commercial structure. The builder generally supplies plans drawn by a professional architect and reflecting the specific features the client desires.

Remodeling or modernizing work is often a good area for beginners. These projects are generally smaller than building a new structure, and it is usually possible to subcontract portions of the work. Still, accurate estimating of time and materials is required, which is often difficult in renovation. Mistakes made or shoddy materials used in the original construction may cost the renovator crew time and expense in making corrections. Because the risks are larger, profits may be calculated higher. Many remodelers calculate prices based on a 20 percent profit margin, compared to the 15 percent generally calculated on a new home.

In remodeling, it is more important to be an organizer than the person who actually does the work. The remodeling contractor must coordinate the schedule for all necessary trades, bringing the workers in and getting the job done quickly, yet to high standards of quality.

The Subcontractor

In recent years a third type of contractor has gained a significant place in the construction industry: the business owner who provides a particular service necessary in the overall construction process.

A subcontractor may perform the electrical work, or install and finish all the drywall, or install the wood trim and custom cabinetry for kitchen and bath. Small operations may be just enough to keep one person busy. A housing development might require that the subcontractor hire and manage a crew of workers.

The contractor pays the sub on completion of the job, which means that subcontractors face the least financial burden in contracting. The remuneration may be determined by competitive bidding for the job, by a calculated hourly rate for the work crew, or by some form of piece rate, for instance, so much per sheet of drywall installed.

Spin-off Businesses

Somehow, home building has a tendency to lead to other business opportunities. Many builders are also land developers, some are real estate or insurance brokers, and still others are involved in commercial building or remodeling. In each of these enterprises it is the honest person with experience, patience, and self-discipline who is likely to succeed.

The majority of contractors got their start by knowing someone in the industry who had faith in their talents and backed them financially. Where do entrepreneurs find such backing? Often at their bank or savings and loan association, where the staff is experienced in dealing with contractors.

Small contractors may trace their start to a single successful job bid. It is important to keep up a steady pace, adjusting to growth when it happens and downsizing if the economy slows. A successful contractor continue to bid jobs in a size category he or she can handle, moving up with experience as a guide.

MARKETING

Of course, it is important to keep your name in front of potential customers. Most construction companies and specialty services rise or fall on the strength of their reputation. Referrals from satisfied customers and a client base from which you generate repeat business are both important. At the job site, your employees represent you and your firm's reputation. They must understand the importance of the attitude they project and the quality of the work they perform.

How do you advertise your business? Some contractors rent exhibit space at home shows or set up a sign with the company logo at the work site. Others hold "open house" at recently completed projects. Businesses take advertising space in telephone directories, are listed in Chamber of Commerce promotional materials, or advertise in the local media. Still, word-of-mouth reports of excellent work are the best and most cost-effective form of advertising.

As you take the first step in your career, study the success of others. You will inevitably learn that luck has little to do with success. Experience, technical knowledge, drive, ambition, and patience, plus the ability to handle stress—all these are necessary if you plan to run your own show.

WEARING MANY HATS

As the business owner, you will be responsible for purchasing tools and materials, planning the work, and supervising workers. Ultimately, the operation of your shop or office is your duty, along with marketing and public relations or advertising.

You will also be the financial manager, deciding whether to purchase or lease equipment. In either case you will handle such details as comparing cost, quality, and features and arranging a delivery schedule and payment plan.

Production planning means deciding how each element of the project is to be built or assembled, by what process, and when. As part of the whole scheme, you must also be able to estimate with some accuracy how long the job will take and who does the work.

If yours is a one-person operation, you already know the answer to the latter. One of the first things a business owner learns is that planning is important. You will be expected to meet deadlines, particularly in the construction business, where contracts often include a penalty clause. Failure to complete work on time could mean having to pay the primary contractor or owner.

A successful business owner learns to anticipate problems and take steps to avoid them. You need someone who knows all the tools and machinery and can troubleshoot before a mechanical breakdown brings work to a halt.

Good supervision is at the heart of quality control. Skill in material handling is vital, since waste can destroy profit. Expensive building materials must be protected from damage or loss at the construction site.

You need to be safety-oriented, for a construction

site can be a dangerous place. As the owner, you are responsible for enforcing employment laws and communicating clearly with workers. In managing your personnel, you must be able to judge character in order to make hiring and firing decisions, assign duties, and protect the good name of your business. In addition, success depends on your ability to motivate workers.

Then there is the matter of running an office. Many entrepreneurs first call on a family member to help out. Often the first employee hired is someone to answer the phone and handle the paperwork. Some record-keeping tasks can be done in the evenings or by taking the books to a part-time accountant. Regularly and accurately, someone must keep accounts up to date, file tax reports, bill customers, pay the company's bills, and write checks to meet the payroll.

With all this work, why would anyone want to own their own business? For many, the lure of being self-employed is strong, even though they work much harder, especially in the early years. Certain people like to make the decisions that affect them; entrepreneurship gives them that control over their fate. Often it is a combination of independence and challenge that motivates them.

Is it for you? Ask yourself whether you would prefer working on your own. Are you willing to take the risks? Luckily, in the construction business you don't have to make a lifelong commitment. If it proves not to be the life you envisioned, you can sell the business and go back to working for someone else.

Support Network

One way to get your new business off to a strong start, and spare yourself some headaches, is to benefit from the experience of someone who is already successful.

Find a mentor who can provide expert advice and

help you sharpen your skills. SCORE, the Service Corps of Retired Executives, operates throughout the U.S. as a sort of "big brother" organization for fledgling business owners.

If you are a woman launching a construction business, you can find support through the National Association of Women in Construction, discussed in Chapter 4. With more than 226 chapters in the U.S. and Canada, NAWIC represents nearly 9,000 women employed in all phases of the industry; it offers networking resources on the latest in construction techniques, women's business ownership, office management, and so on.

Most state governments have come to recognize that small businesses are where job growth occurs. To foster the development of such ventures, many states have established networks of business service centers, offering workshops and information on the nitty-gritty of running a business. Local Chambers of Commerce and Industry also organize programs to assist small-business owners.

Although it is relatively easy to get started as a construction entrepreneur, these ventures involve a certain degree of risk, and the failure rate is high compared to other industries. Still, there are thousands of success stories in communities across the nation. In a few years your story could be one of them.

Energy-efficient windows and doors are installed as part of the effort to return this apartment block to use (courtesy Home Builders Institute).

10

A Day in the Life of...

Of course it is impossible to understand fully what it is like to work in a construction job until you actually *do* the work, but you can get a realistic look through a vignette, or word picture, provided by actual construction workers interviewed for this book.

The emphasis here is on a broad overview of assignments, situations, and experience. You will "meet" several successful blue-collar workers, some of whom are union members, others who are not. Their training and experience reflects the diversity that is typical of today's construction industry.

Our travels will take us to the site of a major construction project—a new international airport. We will also visit a 58-lot subdivision where a custom home builder has crews at work. To see what it is like to prepare for a career in construction, we will interview a student majoring in construction technology.

First, the airport project, where we meet Jamie, and operating engineer. This job involves operating construction machinery including power shovels, cranes, pile drivers, concrete mixers, and concrete pumps.

"I get my assignments through the union hall, but generally I work with the same five- or six-member crew for months at a time. For the past few years, we've been busy from March through the middle of November,

A Day in the Life of...

then have a couple of months off. During the building season there's plenty of work, including overtime hours, so you get a little ahead in savings for the winter months. On projects as big as this one, it's kind of hard to see the big picture at first, but as the months go by you get a feel for how this raw land is being transformed into a finished project. We've moved thousands of square yards of dirt to clear the way for the building that's going up here; it's going to cover four or five city blocks, what with all the levels and concourses. Then there are the runways and hangars, all at various grades. Building an airport is like several big projects wrapped up in one. There's road building, with all the access lanes and parking lots on one side of the terminal, and then the runways and taxi lanes on the business side, where the planes are located.

"I've been at this for fifteen years now, running everything from a backhoe to a grader and the big dozers. I got my training in the military, but most of the guys have been through the union apprenticeship. A few went to trade school, including the Job Corps. You need to see well and have good coordination to run one of these machines. The noise and vibration are problems sometimes, but now we have to wear ear protection. Oh, yes, and the sunblock. All this business about the hole in the ozone. You get a lot of sun in this work, so you have to protect your skin more than before.

"Here's a skill you can take anywhere in the country. If there's one thing I miss, it's a summer vacation, but when my neighbors are digging out to go to work on a snowy winter morning, I figure that's my pay-back time. If there's one thing that bothers me, it's being away from home during the week. Take this job. I live nearly three hours away—too far to commute daily. Road-building assignments are even worse. On an interstate project, you could be in another state for months at a time. If it ever gets to me, I could start an excavation company of my own, but that would mean smaller projects. I'm looking forward to the day I can drive through this place with my kids and show them the work I had a part in."

At the terminal area, workers from the plumbers and pipefitters union are using power-driven machinery to extract core borings. These columns of soil brought up from well below the surface help to determine the structure and quality of soils that will support the weight of the building. A common activity at building sites of any size, the work is often assigned to second- or third-year apprentices.

"We bring up the samples, and turn them over to the engineers. I've done this for hotels, office complexes, shopping malls, and two school buildings, just this year," one of the workers explains.

"I've been stringing water pipes, too, making connections and testing for leaks. Two afternoons a week, the apprentices knock off at noon and go to school at the Pipe Trades Center. Classes take 144 hours a year, so it's a change of pace. I don't care much for reading, but we have demonstrations and labs, mostly.

"I'm interested in what they call 'large work,' public works projects, dams and flood control, that sort of thing. I'd also like to move South, where there's work all winter long. The union is definitely the way to go, as I see it. The pay is good, and so long as you're careful and work with careful people, accidents are not a concern for me. I've lived in the same small town all my life. I'm ready to travel and see some of the country. Eighteen months from now, I plan to pick up my journeyworker's card and take off."

Less than half an hour north of the new airport by commuter flight, we get an aerial view of a housing subdivision under development. Leafy treetops mark the buffer zone between the curving residential streets and a connector highway. At ground level, we hear the ring of hammers and whine of power saws. Our first stop is at the office of the developer, a **custom home builder** who has just finished an evening meeting with a couple planning their dream house on a cul de sac two streets over.

"This was my first chance to see their drawings, and it's a nice house; it will fit right in with this neighborhood and be relatively easy to build. They haven't made all the decisions about materials to use inside, but that's all right at this stage of the game. Better that people take their time and be satisfied right from the start, than change things as the work goes along . . . too expensive then, for them and for me.

"Tomorrow I'll talk with the guys who clear lots and excavate our foundations. These homeowners are very specific about which trees they want to save and how they want some other details of the earth-moving handled. It's my job to pass along their ideas and see that they are respected. I do that in two ways; I talk with the workers, and I give the subcontractor a written checklist, keeping a signed copy for my own files.

"Throughout the job I'll be on hand for every major event, checking delivery of the building materials I've ordered, keeping an eye on time schedules and quality control, making sure the building site is kept clean and safe during the day and secure at night.

"Oh, yes, there are people out there who would love to make off with all these valuable building materials. Once, back in my early days, someone carried off all the materials for a 12-by-30-foot outdoor deck. Because it was a package deal, all the treated lumber was stamped with my order number, something the rascals didn't notice in the darkness. The sheriff had my invoice with the order number on it and recovered all the lumber the next day.

"Probably the most important part of doing this job well is to enjoy working with different people. The clients are great because the house they are building is their dream. The staff at the savings and loan that provides the construction funds are people I've known and worked with for years. We trust each other. I also try hard to do business with subcontractors who really like the work they do. I keep delays to a minimum, especially their pay-outs, and in return they see that I get quality work and meet my deadlines. It's a two-way street.

"You have to keep up with the latest materials and technology to bid contracts competitively. I do that by making the rounds of trade shows during the winter and keeping up with several industry publications. My membership in the National Association of Home Builders is a great help too."

Six weeks have passed, and the builder has introduced us to a **finish carpenter** at work in the home that was little more than an architectural drawing on our last visit. Now the house stands stately and tall, surrounded by trees as the owners intended. The carpenter is inside, unpacking custom-built kitchen cabinets from their shipping cartons. He wears safety glasses and running shoes. A leather belt with pouches holds assorted hand tools, and a hammer hangs from a metal loop at the right side.

"As far as I can see, the finish carpenters have the most fun building a house. We get to work with the best materials, the smooth, milled lumber, and we handle fine cabinetry like this. Rough carpenters, sometimes called framers, build stud walls and roof trusses, install

A Day in the Life of...

the decking, and build forms for concrete. I've done it, but it's not my idea of fulfillment. I've never liked the idea of having my hard work covered up at the end of the job. What I like most about finish work is the precision it demands. There's little margin for error. Of course, we also hang the doors and case the windows, but those are still finishing touches that show. I take pride in this work, and that makes for a better job.

"The tool I depend on most? No question about it—it's the power miter saw. You want a clean, smooth fit wherever lumber meets lumber at an angle. When you're making maybe 100 measurements in a day and cutting as many pieces of lumber, you want to do it quickly and efficiently. That's why the power miter goes with me on every job. I handle it carefully too, to make sure it's never knocked out of alignment. I have a set of wood chisels I inherited from my father, who did some cabinetmaking in his time. I value them a lot. Even with preassembled doorframes and trim components, there are times when you have to shave down an edge or finetune something for the best fit.

"Something that I used to do a lot and now is coming back into style is the hardwood floor. The width of the flooring and the finish treatment change with interior design trends, but the method of installing these floors stays the same. It's time-consuming, sure, but the floor will look great as long as the house stands. Few other materials last that long. I've even installed wood floors in kitchens and baths over the past few years. This is a veneer product, thin strips of oak bonded to a resilient base and topped by a thick polymer coating. The natural wood shows through, and the shine looks like the real thing, only better—water and spills can't touch it. When something new comes along and it proves to be a good product, I like adding it to my bag of tricks."

The **electrician** we meet has been a journeyworker

for a year but has been on the job just six months. He learned the trade during a four-year hitch in the U.S. Navy, and he previously worked on an aircraft carrier several times longer than the high school football team in this, his hometown.

All electricians are trained to follow National Electrical Code specifications, whether they learn the trade in the military or through apprenticeship. After taking the IBEW union entrance exam, he was issued a union card and reported for work the next day with a contractor.

"I keep in touch with some of my Navy buddies, so I know that very few of them got good jobs as civilians as quickly as I did, or started at such good pay. I just got my second raise, and I'm at $21 an hour already. The industry average is just over $26 an hour, plus fringe benefits.

"Of course, in construction you have to take into account the seasonal nature of the work. I learned how to save money in the Navy, so I'm not worried about the chance of a winter layoff. If it looks like a slowdown is on the horizon, I can work some overtime and save that money for later.

"When I started looking around for a career, it didn't take long to focus on being an electrician. Two of my uncles are journeyworkers, and my family is in construction, so I've been around building sites all my life. The real surprise was when I met a Navy recruiter at our high school career day and discovered that I could learn the trade while I got to see the world. Our homeport was San Diego, but during my hitch we had a cruise to the Mediterranean and took part in training exercises in the Pacific. To get shore leave in Hawaii, Rome, and Spain all in the same year—who would have believed it?

"I don't want the paperwork hassles and re-

A Day in the Life of...

sponsibility of being my own boss. Not many electricians are self-employed, and there's not a lot of job turnover in the industry. To me, those two facts mean there must be a lot of employers out there who make an effort to keep their electricians happy.

"The union and contractors are into a lot of new programs to recruit nontraditional workers, especially women, blacks, and Hispanics. Our company is supposed to get two new apprentices the first of the year. It will be interesting to see who they are. When I took my journeyworker's exam, there were people across the hall taking the apprenticeship test. It was a pretty mixed group of folks. I say, if they're good at what they do, more power to them.

"A lot of construction jobs require that you buy a set of expensive tools. An electrician gets off easy in that department. You use a couple of different tools on the job, but the only ones you buy are things like screwdrivers, pliers, hacksaws, and utility knives. The company provides pipe threaders, conduit benders, power tools, and test meters, but of course, you're expected to take good care of them.

"One thing you need to know—an electrician stands up to do a lot of the work, often in cramped spaces. If you have bad arches, or get claustrophobic, give that some thought. Something else, you have to have good color vision. The color-blind electrician has a real problem, since wiring manufacturers use a system of color-coded wires that must be matched up and wired together.

"I like to work with people who are as safety-conscious as I am. If the people around you are careless, you could fall or get hit by a falling object, not to mention the danger of getting a shock."

In the past six months this electrician has worked with a four-person crew wiring an addition to a nursing home, and also worked alone installing dozens of dishwashers, hot water heaters, and other appliances in the new homes of a subdivision. He has also installed underwater lighting for several swimming pools.

"If you enjoy solving math and science problems, then understanding basic electrical theory should be a challenge you can succeed at.

"I like to do both the rough and finish wiring in new houses. To me, it's interesting to see how a project has taken shape since the first time you worked on it. It's a thrill to drive through a neighborhood a year later and realize that your work made possible all those lights glowing in the windows of the houses there."

Drywall mechanics transform the interiors of buildings under construction from skeletal frameworks to efficient systems of enclosed interior spaces. To do so, they measure and install large sheets of gypsum-backed plasterboard, usually with a pneumatic tool, creating sturdy walls and ceilings. Then the seams are carefully taped and finished with joint compound to make the adjacent panels look like one smooth surface. In the course of a day a drywall crew may erect scaffolding and perform work at some height. Some drywall finishers even wear specially designed stilts while they work, to reach overhead areas with ease. It is true that the drywall installer lifts heavy materials, but he or she rarely labors alone. Typically, a team works together to hang the drywall, then splits the rest of the work, one taping and applying joint compound to the seams while the other does the finishing details of sanding and smoothing.

The drywall finisher must mix materials in proper consistencies and be skillful in the use of various trowels and finishing tools. Skill and experience bring a "good eye" for determining whether a given drywall will dry to

A Day in the Life of...

a flat, smooth surface. In certain kinds of finishing treatments, ceilings and sometimes walls are finished in a textured surface, which the drywall crew applies with special tools.

Our look at the career of a drywall mechanic begins at the attractively landscaped entrance to a housing development. Around that bend is a house under construction, and inside is the worker we seek. A painter's cap, bib overalls, a paper mask, and goggles distinguish him. As we shake hands, we notice the gritty coating of plaster dust.

"The sign of a true professional," he says, laughing. The developer got him started hanging drywall three summers ago.

"I'd been working out at the gym during the off-season, but with summer coming I wanted a job. When my football coach said he knew a way I could do some daily weight training and earn money for it, sure, I was interested. Turns out the weights I got to lift were these drywall sheets. Still, I like the idea of building walls where there were none before. I like making the panels fit tightly. Believe it or not, we don't use hammers. This power impact tool drives drywall screws through the panel and into the studs. The work goes quickly, too.

"As high school graduation grew nearer, I realized that college was not what I wanted. I wasn't interested enough in any subject to study it for four years, so I came back to work in the spring, afternoons and weekends first, then full-time after school was out. The developer offered me a chance to become one of the subcontractors here. You can bet that made me one proud 19-year-old. It's regular, predictable work, and I don't have too many worries about bookkeeping and business details—not yet, anyway. Sometimes if the painters are really busy I stick around and paint the rooms I've just built walls and ceilings in. I can work all year, since the heat and electrical work are in by the time I get started on a house. Sometimes I work solo, but it's never lonely. There are usually plumbers and electricians working around somewhere. Most of them are independents too, and we eat lunch together most days.

"The future? I'm already making $12.50 an hour, after you subtract the cost of my health insurance. There's a lot to be said for liking the work you do, and I still can't think of a college subject I like that much."

The sound of a dump truck outside signals that it's time to move on. Painted as green as an oak tree in summer, the truck carries the name of a landscape contractor on the door panel. At the wheel is our next interview subject, a working supervisor with the landscape contractor.

We follow the truck down winding Tamarack Lane to the site of a contemporary home now nearly completed. As the dump truck backs into the drive, we can see it is loaded with rolls of turf.

"Instant lawn," the **landscaper** explains, removing a dirt-stained glove so we can shake hands. "We don't always do a lawn in a day like this. Farm-grown turf is expensive; most people have enough time before moving in to let us prepare the ground and seed it. That way nature does the work. A business executive who's coming here from another state just bought this house. The family have four small children, and they'll be here next week. Turf is really the only practical option here."

In addition to contracts for much of this subdivision, the landscaper supervises contract maintenance for the grounds of several office complexes, a utility company, and a golf resort now under construction. Each job means supervising a crew of two or more laborers, plus

A Day in the Life of...

meetings or telephone contacts with people from the individual companies.

"Most days I can wear jeans to work, but when there's a meeting I dress up more, maybe even wear a tie if it's an important new account. Luckily for me, the partners who own the company handle most of the meetings, but I do spend some time behind a desk. I prepare the invoices, order materials, make up the work schedules, that sort of thing. The best part for me is still looking at a landscape you just planted and visualizing how it will be in the different phases of maturing. What you created is better than what was there before, and you know it will be there for people to enjoy in the years ahead.

"Sometimes if we're short-handed, I go out with one of the crews and pitch in. That's something I miss a little. Moving up to supervisor meant more green in the pay envelope, but I wouldn't want to miss out completely on the chance to get my hands in the dirt. That's why I wanted to keep the 'working' part in my job title."

In the course of our conversation, we learn that this landscaper got into the business straight out of high school after having worked two summers for the same company. Those summers were spent on a crew doing contract maintenance for a natural gas company. Like many utility companies, this one follows a schedule of repacing older gas lines in the summer months, a job that requires a backhoe to uncover the lines and a landscape crew to rake or roll the filled-in trench line, then seed the disturbed area with grass and dress the surface with straw. Once the grass seed gets established, someone comes back to rake away the straw mulch.

"Nothing to it. I made $5 an hour, and when it rained we did some light maintenance or worked around the nursery. I knew the company did other kinds of landscaping work, more challenging stuff, and I made up my mind to establish a good record and get myself a shot at one of those jobs. That was six years ago. Now I'm glad I did. There's just enough math and reading blueprints to keep me on my toes, and of course, you have to be diplomatic with the clients, but most of that stuff the partners take care of. I keep an eye on my crew and the equipment. I still get time to stop and smell the roses, as they say. I'll never be a millionaire, but on the other hand, I'll probably never get an ulcer, either."

"When I told my dad I wanted to study construction technology in college, he was all for it. It was my granddad, the one who founded the family business, who objected. 'What can they teach you that you couldn't learn on the job'—that's what he wanted to know. Well, I hate to hurt Granddad's feelings, but he doesn't understand computers. In fact, he doesn't even want to understand. This is the future, though, along with all the new materials. That's what I'm really here for."

"Here" being the campus of a state-related technical college where the accent is on practical education. Students earn associate degrees in a variety of fields, from computer repair to hotel management, though the focus has traditionally been on the building trades.

"One of these days, I'll be managing a building crew of my own. I don't want the employees questioning my ability, thinking I'm supervising them because of my last name, not my skills. Besides, there's a lot of business training involved in keeping a company going. I need more math and computer training than you get even as a vo-tech carpentry student.

"I'm really interested in the design/build concept, where the construction company gets involved early in the process, not just interpreting somebody else's drawings for a project. There are two ways to get started in design/build—have twenty years' experience or some technical background. I'm taking the education route. Right now, our company subcontracts all the excavation and foundation work. I'm not saying that I want to handle all of that work, but we could expand and get a better handle on scheduling if that were part of the package we offered clients. I'm also interested in new materials. With lumber prices the way they are, steel and high-tech plastics are beginning to be competitive. Working every day, my father and granddad don't have time to attend workshops and seminars. These are areas of the business where I could carve out a niche for myself, make a real contribution. Mostly, I'm here because of that competency issue. I want to show people I know what I'm doing. In this program, you have to declare a specialty halfway through the course. It's either carpentry or home remodeling. The carpentry focus takes in concrete and commercial construction. Remodeling leans more toward plumbing and electrical work for the trades. I'm not so interested in the theory that goes with electrical work.

"Putting together accurate bids that reflect time and materials in a job—that's what matters when you run your own business. There are twenty-three employees counting on the bid. Computers and a knowledge of materials make that work more scientific. Dad knows that's the wave of the future in contracting. I guess that's why he's willing to have me leave the business for two years and get this training. He wants to see the company keep going into a new century.

While this individual was essentially "born" to the business, our next interview subject was introduced to painting and decorating almost by accident.

The first thing we notice about our next interview

subject, a **painter/decorator**, is the length of neon-pink yarn holding her hair back in a jaunty ponytail. Dressed in pale denim bib overalls with just a few paint splotches, she wears a bright orange tank top, a painter's cap, and red high-top sneakers.

"Not your typical company president's wardrobe," she points out, "but all the same, that's who I am." How did a former office worker and mother of two launch her own successful contracting business?

"I hate to admit this, but it was almost by accident. I was between clerical jobs and had just signed up with a temporary service when my mother asked if I'd be interested in painting the fence around a small apartment building she owns. I thought, "How hard could that be?" About the third day I was out there, a woman called to me from the balcony of the building across the back alley. Turns out she needed some rooms painted, but as a widow living alone she was reluctant to call strange men into her home.

"That was my second job, and before it was done the lady had rounded up three or four more in the same apartment building. Then the manager of another apartment building, also a widow, called me to bid on more painting jobs. I got into my first wall-covering work in that building too. I sort of exaggerated my experience, then dashed off to the library and a home center for research. One let me borrow an instructional videotape, and I learned a lot from watching it a couple of times. Since I had two weeks before the job began, I decided to experiment on my daughter's room, which turned out really well. What a confidence builder! When it came time to start the job I discovered that I could dip the goods [sections of wall covering] into the water box and book [loosely fold] them like a pro. Luckily, the pattern was easy to match and there weren't many tricky details to cut around."

Painters may work from a ladder, scaffold, or hydraulically operated crane bucket, as illustrated here (courtesy IBPAT).

How did she handle the business details in the early days? "Well, my husband is a building contractor, so he could advise me at first. I was plenty scared about the first bids I made, but even then I never lost money. Making a good estimate is not as easy as it may sound. You have to understand how much time it will take to get to the job, set up, and get the work done. Then you have to make sure that what you budget for materials is realistic. I get a contractor's discount now, which sure helps, and most of my suppliers are willing to bill me for 30 days. There are few jobs you can't finish in that time.

"After those first few jobs I got business cards printed up and asked my customers for referrals. It didn't take much working capital; in fact, I used the Christmas club money I'd been saving on my last office job to buy drop cloths, brushes, and an extension ladder. I used the phone at home but got an answering machine for the daytime. A business like this can grow quite a bit and still operate from home, which is important, because controlling overhead costs could mean the difference between failure and success. For the first couple of years I did everything. I was the job estimator, the bookkeeper, the purchasing agent, and the painter too.

"In the early days you wield a paintbrush too, and pay yourself an hourly wage, plus a percentage for supervising if you have other painters on the job. You also charge a sum over and above the actual hourly rate paid to each member of your crew, maybe 15 percent. Part of that is profit, and part is needed to pay the costs of operating your business. You may also decide to build some profit into the cost of materials, or include a charge for your time and equipment used in transporting materials to the job.

"Most of the people I work for are at home while I'm there working, and we get to know each other a little. Getting paid for a job has never been a problem for me, although I've heard from other kinds of contractors that they have trouble in that regard. Maybe it's because of the personal contact. Some of my clients have told me they went for my bid specifically because I was a woman. For the elderly ones, it's probably a personal security thing. They just don't feel safe letting strangers into their homes. I guess others see me as someone who won't make an enormous mess while doing the work. I

As basic requirements, a painter should have no fear of heights and be free of allergies, particularly to chemicals and solvents. A sense of color and attention to detail are also attributes of the successful painter (courtesy IBPAT).

realize that is an advantage, so I take special pains to keep things tidy. In the winter, when I may have to track in and out with materials, I provide a vinyl runner for the floor or carpets from the outside door I'm using to the room where I'm working. I also go over the room meticulously when it's done, making sure there are no little adhesive marks anywhere. I even offer to rehang the pictures and wall ornaments. Those are little things, but people appreciate them."

Asked how someone just entering the working years could decide whether painting and decorating is a career they might pursue, our subject ticked off a list of questions: Do you have good manual dexterity? A good sense of color? No fear of heights, or allergies?

If so, she suggested that the skills taught in an apprentice or vocational program, combined with basic business knowledge, can help you become your own boss in less time than it takes to earn a college degree.

A pro needs to handle both brush and roller expertly. You can learn those techniques in trade school classes. There's a saying among painters that a good brush is as important to a fine job as the paint or varnish. Watch how the pros care for these tools of their trade, she advised, adding, "A good painter knows what a good brush is—containing the highest-quality bristles with a properly tapered edge. Such a brush must have proper balance and a comfortable handle." A painter also handles a roller expertly. Two designs, the dip and the fountain pressure types, have been in use since the end of World War II. A good painter must also master a paint sprayer, which makes it possible to cover six times the surface done by a brush in the same time. However, use of a sprayer requires more preparation time for masking and taping off areas that are not to be painted. A sprayer handles many lacquered and high-gloss enamel surfaces better, and with less material.

A Day in the Life of...

"At one time or another, most people have painted a bedroom or kitchen. With today's prepasted wall coverings, a lot of people have experience in hanging paper too. Painting and paperhanging for a living are much the same, but on a larger scale."

What, exactly, does the work consist of? Let's talk about painting first. You need to prepare the surface by sanding or scraping away old paint. Good old-fashioned elbow grease may be necessary to clean off dirt, dust, grease, or loose paint. In some cases the painter uses chemicals, electric scrapers, sandblasters, or blowtorches to remove old paint. Next you fill nail holes and cracks in plaster, synthetic materials, metal, concrete, stucco, and wood. Both old and new surfaces may require a first coat of primer or sealer, especially if you'll be covering a dark shade with something lighter. The final surface covering is applied in two coats. The first one seals the surface and keeps succeeding coats from penetrating. The second and any additional coats contain the color and provide resistance to weather and wear.

The job may require ready-mix paints, or the painter may blend and mix colors and additives into a base paint to achieve a custom tint and consistency. You need a basic knowledge of such products as paint thinners, drying agents, and fire or mildew retardants. Different jobs require different blends, and it is the professional's job to know what kind of coating is best for the material involved. Common surfaces include plaster, drywall, cinder block, concrete, brick, wood, metal, even adobe, a sun-dried block native to the Southwest. The skilled professional knows which tool is best for each individual job.

The professional paperhanger, too, is concerned with a smooth work surface at the start of a job. If the room has been papered before, it may be necessary to remove old paper by soaking and steaming it. Cracks and holes

in the plaster must be patched and smoothed. On new plaster, alkaline spots must be neutralized with a special solution. Careful measurements are required to achieve a professional look and avoid wasting expensive materials. Mixing and applying adhesives comes before careful positioning of the wall covering, which is then smoothed with brush or knife. Patterns must be carefully matched to assure that the design continues from strip to strip without a visual break.

"My story is a little unusual, in that I had never worked for a contractor before starting out on my own. I read somewhere that about 80 percent of all painters and decorators work for a contractor. The rest are independents. Some specialize in new homes, apartments, and office buildings. Others concentrate on repair, remodeling, or updating projects such as scheduled maintenance for hotels and office complexes. Some of those places go through an entire building every five years or so, repainting or replacing all the wall coverings.

"I've met some people who are union painters, members of the Brotherhood of Painters and Allied Trades. Union wages are a lot higher than I charge, and from what I hear, the best-paid painters and decorators work in California. That's not for me, though. I don't want to leave town, and I actually like to work alone most of the time. Now that we can afford family vacations, I like being able to coordinate my schedule with my husband's time off and the girls' school breaks.

"Once you get established, you have opportunities to supply wall coverings direct from the factory to your customers. That aspect of the job offers another means of making some money, since you get deep discounts

Many professional painters expand their range of skills by learning to install wall coverings as well (courtesy IBPAT).

and can keep a commission on the sale the way a store would.

"Lately I've become fascinated with the more artistic aspects of the business. It started with stenciling, but now I'm into some pretty exotic special effects. You can come up with some amazing textures and finishes with a little study of the color palette and the light-reflecting qualities of paints. For instance, I really enjoy a technique known as marbleizing, where you use a turkey feather to create the look of veins of color against a darker background. It can be amazingly authentic-looking, and I feel like an artist when customers show off my work. I've done a couple of fireplaces. This winter I'm planning to do all the fireplaces in my own home."

Asked whether she had encountered any unexpected or negative aspects of the job, our subject said that working in hot summer temperatures took some adjusting. "Now I carry a big water bottle at all times and a thermos filled with one of those sports drinks that keep your electrolytes balanced. The trick is to avoid dehydration. Through perspiration, you can lose a couple quarts of body fluid in a short time. Since you could be working on a scaffold you want to avoid getting dizzy from the heat."

The star of our next success story, a **plasterer**, began with a background in high school art classes, yet could not find work that allowed her to use those skills to earn a living. In fact, despite a diligent job search, she seemed to be caught in a frustrating, and increasingly frightening, downward spiral. "My dream of becoming an artist vanished. I had no job, no income, and I was staying in a women's shelter."

At a government employment office she saw a poster

advertising free Job Corps training classes. After finding out more about the program, she filed an application on the spot.

Although months of hard work and dedication were required, after mastering the basic aspects of plastering, our subject had a chance to study ornamental sculpting, a specialty that let her artistic talent shine. One of the most rewarding courses focused on scarfito plaster carving, a specialty technique. As an outstanding student, she was selected to represent the Cascades Job Corps Center in Washington State during festivities marking the Job Corps' 25th anniversary, an event held in the nation's capital.

"While I was there, Spencer Christian of 'Good Morning America' interviewed me as I demonstrated the technique of scarfito plaster carving. Having my plastering skills broadcast on national television was definitely an interesting and rewarding experience."

Plaster, more than perhaps any other building material, lends itself to creativity in design. Skilled, talented people can "put their mark" on a building by creating such details as plaster medallions and cornices. These are rare talents that until recently few experts were passing along to the next generation of plasterers.

In addition to the opportunity to express her creative side, this worker saw other advantages in her vocational choice. "I chose plastering for several reasons: the excellent wages, the job opportunities after graduation, and the variety of skills learned . . ."

Her Job Corps experience paved the way to a bright future, where lifelong learning certainly plays a part. "I'm now in El Paso, where I have decided to attend college this fall at the University of Texas. I'm interested, of course, in the arts as my major, specifically sculpting. Thanks to Job Corps and the plastering trade, I know I'm headed in the right direction."

From the ranks of nearly 40,000 **apprentice carpenters** in the United Brotherhood of Carpenters and Joiners, 40 have made the journey to a convention center in Chicago where "the contest" is under way. The contest is an annual event in which the best and the brightest carpenter apprentices compete against the clock to show what they know about building.

Each apprentice must work alone with his or her own hand tools. Contestants are given a set of plans, a package of materials, and a certain number of hours to complete the work. Projects are chosen at random, but in last year's contest they included installing a small bay window, building a porch, framing and enclosing a small addition, and installing a set of stairs.

The contestant we are watching works for the custom home builder we met at the beginning of this chapter. He is experienced in metal framing and building forms, which will be to his advantage, since blueprints for the small addition are in his contest packet.

At first the work areas are quiet, as contestants examine their plans, check materials, and assemble their tools. Soon the air is full of the whine of power saws.

The panel of judges walk slowly back and forth along the row of projects, using a comprehensive checklist to determine point values for each stage of each project. In addition to demonstrating craft and technique, contestants are expected to pay attention to the details. For instance, is the workplace safe and tidy? Judges watch that aspect of each job.

During the nearly three years this apprentice has been learning carpentry, he has been given assignments with increasing responsibility and challenge. One evening each week during the building season and three nights a week during the winter when the work schedule was less hectic, he has attended classes sponsored by the union and designed to support his on-the-job training.

Apprentice carpenters on a housing rehabilitation job site demonstrate the philosophy of a careful worker: measure twice, cut once (courtesy Home Builders Institute).

"We're getting near the end of the classes now, which means we're studying things like time and material management. On the craft side, we're building and installing some of the new custom components for kitchen cabinets, top of the line items like rolling shelves and bins fitted inside the cabinet space. Last week I put together a spring-loaded shelf designed to store a heavy food mixer out of sight, yet raise it to a comfortable working height at the touch of a button."

The finished product, a new home, represents the work of more than a dozen skilled construction workers, from the equipment operators who excavated the foundation to the landscapers who put the finishing touches on the lot (courtesy Armstrong World Industries).

Asked whether he had any words of wisdom to offer people considering a career in the construction trades, he replied, "Measure twice, but cut once," a bit of advice that has helped many carpenters become meticulous and skilled journeyworkers.

The apprentice carpenters will be busy with their contest entries for most of the day. There are prizes in several categories, but even more important to the

A Day in the Life of...

contestants is knowing that their work and talents have been on display for some of the most successful employers in the nation. By day's end many of the contestants have received multiple job offers.

Of course, these brief word pictures cannot truly convey the reality of a career. Like all jobs, there are good days and bad days. If you become a skilled, reliable worker with a trade that is in demand, the balance of good days will be heavily weighted in your favor.

As America moves closer to a service-based economy, there will be fewer opportunities for workers to pause at the end of a job well done and say, " I built that." If that feeling of accomplishment in a tangible product is important to you, then you should carefully consider carving your niche in the construction trades.

11

Toward the Future

Striking that delicate balance between time and materials has always been a critical factor in the success of the construction industry. These key elements, along with the impact of skilled-labor shortages in the future, will form the challenges for a new century. Of course, new construction techniques will be developed, along with many new tools and building products and innovations in how builders organize their work. These factors will have an impact on your training for a construction career, and how you pursue that career.

Making the building process more efficient is one way in which the construction industry is boosting productivity and keeping its services affordable. A prime example is the concept known as production engineering management, which means delivery of materials to the construction site in a way that is both practical and cost-efficient, requiring less labor time in handling and storage, with less chance of damage to materials or delays at the job site when materials are not there when needed.

Many volume builders are familiar with a problem-solving and planning technique called the Critical Path Method, which produces a logical, precise construction schedule. Great improvements have also been made in materials handling and packaging. Structural parts are

Toward the Future

stacked or wrapped or taped in such a way that each is available as needed, from first to last.

Entire system packages can be moved by materials handling machinery, from the delivery truck to the spot where they are needed. Whole sections of a structure are commonly erected in this manner, as builders have turned from human to mechanical laborers. Depending on the specific job to be done, builders rely on forklift trucks, carryalls, bulldozers, tractors, graders, power shovels, backhoes, trenching wheels, and drilling and hole-boring equipment.

Increasingly they call upon electric saws, drills, staplers, routers, planers, nailing machines, electric power mixers, and powerful spray guns that distribute concrete and apply paint. Small portable lifts and chain saws are also common sights at today's construction projects.

Research for the Future
Let's take a closer look at how the construction industry, particularly homebuilding, is preparing for these challenges.

IBACOS Inc. is an alliance of leading companies in the homebuilding industry dedicated to research and development of methods to keep homes affordable and energy-efficient. The name is an acronym for Integrated Building and Construction Solutions. This unique consortium of leading companies in the building industry is developing better products, building materials, and construction processes and taking their solutions into the industry faster than ever before.

These industry leaders are working together to meet the complex issues facing their industry—specifically, affordability, skilled labor shortages, low productivity, environmental pressures, energy concerns, and changing American lifestyles and demographics.

Much of their effort focuses on building a series of test houses under research conditions. Affordable and of high quality, these homes are also environmentally responsive, quick to build, and adaptable to changing lifestyles. The new products, systems, and processes that go into the IBACOS test homes will ultimately help buildings meet the needs of homeowners in the next century. Not only are the test homes built with the latest products and technology; they are also carefully monitored for energy performance.

IBACOS, Inc. has five "core" members with the most intensive involvement. They include G.E. Plastics, the world's leading supplier of engineering thermoplastics; Burt Hill Kosar Rittelman Associates, experts in architecture, engineering, and energy systems; Masco Corp., a family of companies specializing in kitchen, bath, and furniture systems; Ryland Homes, one of the nation's largest home builders; and USG Corp., a leading producer of engineered gypsum materials and building products. Other active partners produce electronic control systems, fiber optics, air-conditioning products, and major appliances. As you can see, all of these partners have a stake in the future of homebuilding in America.

Since 1991 this consortium has been amassing knowledge about the changing needs of builders and homeowners. The partners are developing products and systems for a home's structural shell, the kitchen and baths, indoor air quality, and electronic home systems management. By collectively examining how materials, design, engineering, manufacturing, and delivery can work more efficiently together, the IBACOS approach is improving many stages of the complex process of building homes.

According to IBACOS research, the basic technical

concepts in building affordable, efficient homes for the future are three:

Disentangling—separating the structural shell of the house from its interior and service systems.
Core Systems—integrated, engineered building components and systems for use in new construction or remodeling. These "open" systems include the shell (foundation, wall, roof); interior partitionings; kitchen and bath; power, signal and control; plumbing; heating, cooling, ventilation; indoor air quality, and service distribution chases.
System Integration—Making the core systems work together efficiently in a total home environment.

Michael Dickens, president of IBACOS, Inc., calls these core systems the key to the entire process. "We describe the core systems as a kit of parts designed and coordinated by computer, manufactured by member companies, and capable of being used in a wide variety of ways in both new and old buildings. Core systems are designed for flexibility, customization, quick and efficient assembly, and streamlined delivery to the construction site."

As the first step in demonstrating the IBACOS philosophy, the alliance developed full-sized sections of a house shell, interior partitioning, and kitchen and bath areas as its Innovation Center in Pittsburgh. The alliance has also constructed two research lab homes in a Pittsburgh community, where product monitoring and tests are now being carried on. The next phase of the project, targeted for 1995, is to construct an entire community of demonstration and prototype homes, to be occupied by families living a typical American lifestyle.

Because the federal government is also concerned

about the affordability of housing in America, the U.S. Department of Energy has partially funded the research lab homes in Pittsburgh. Innovative projects such as IBACOS will ensure the future of America's construction industry.

FACTORY-BUILT COMPONENTS

Whereas the Pittsburgh project is major in scope, a number of equally important research and development efforts have been yielding important results on a smaller scale. Take the concept of factory-built components, for instance.

Such manufactured components are making inroads in the construction process. Consider, for instance, the wall system described as a "stress-skin panel." The product of a factory assembly line, this energy-efficient item consists of a thick (four inches is typical), rigid foam panel sandwiched between a weather-resistant exterior sheathing material and an interior layer of gypsum wallboard. Produced in 4 foot by 8 foot panels, they are stacked in a truck and lifted into place at the job site by crane. Properly handled and installed, the panels are rugged, energy-efficient, and real time savers.

Ironically, stress-skin panels were initially developed for use in constructing contemporary homes based on the centuries-old post-and-beam style of architecture. The home draws its structural strength from an interlocking system of posts and beams, so the walls are not required to carry the weight of the roof.

These innovations do not mean the demise of the "stick-built" home, by any means. Because the economy of factory-built structures is directly related to the distance between factory and building site, experts in the field of homebuilding do not foresee the end of the new home constructed entirely on its foundation.

Steel-framed homes are also coming into their own,

Toward the Future

providing another opportunity to use modern engineered roof and wall systems. Builders of tomorrow's homes, shops, offices, and factories will be incorporating new products, materials, and methods as innovation brings them to the marketplace.

A new era of building materials made of structural polymers—that's right, *plastics*—are just beginning to come onto the market, the result of years of intensive research and development. Even before there was IBACOS, General Electric Plastics Division was building industry alliances and combining research efforts to find innovations in homebuilding. At least fifty U.S. and international companies joined forces with G.E. Plastics Division in developing innovative new materials and building techniques, which began field testing in 1991 at the G.E. Plastics House built near Pittsfield, Massachusetts. Michael Dickens, now head of IBACOS, Inc., was the leader of the New England project, too. These products are expected to have a big impact on how structures are built well into the twenty-first century.

But what about wood? Since the dawn of time, wood has been the primary building material, and it is likely to remain so, considering efforts that have been made to conserve and harvest timber. Even buildings to be constructed of poured concrete require wooden forms, framing, and finishing products such as milled woodwork and cabinetry.

Science and technology have been applied to product development in the forest. Of recent vintage is Oriented Strand Board, a sheet product with great strength and durability, manufactured from what was once sawmill waste. Rugged glue-laminated beams are formed under pressure from a stack of milled planks coated with industrial strength adhesives, providing the load-bearing strength of a solid log—perhaps more.

Glue-laminations also permit the production of lumber in large sizes and shapes that cannot be cut freely at most lumber mills. Today imposing and beautiful glue-lam arches vault the ceilings of large cathedrals and public auditoriums where once the choice of building materials would have been limited to reinforced concrete or structural steel.

Strong new adhesives are taking the place of tacks and nails, performing well in a range of temperatures and even under water. Chemical foams serve as insulation, and a new generation of solar products is on the horizon. Technology has also improved traditional products for heating, cooling, and ventilation in homes and other buildings.

The Sky's the Limit

Many of today's young adults were born about the time an American astronaut left the first human footprints on the surface of the moon. In the decades since, exploration has increased the prospects of building one or more permanent space stations to orbit the earth. While of prefab design, the components of a space station will still require skilled hands for assembly. In your lifetime it is likely that space shuttle crews will encompass construction mechanics as well as scientists and astronauts, so be prepared to turn in your hard hat for a space helmet.

Back on this planet, a number of advances in building technology will involve electricians, who will be required to add electronics to their job skills.

Already available with today's technology is a network system that enhances the electric, telephone, and video capabilities of home wiring. Installed as a central hub, one system permits the home telephone to handle up to four incoming lines, assigning the calls to any phone in the house. Telephone service

Toward the Future

can be further customized by transferring calls, putting callers "on hold," and accommodating conference calls, intercom uses, and room-to-room dialing. Coaxial cable permits cable TV and VCR signals to be directed to any room. With the addition of closed-circuit TV cameras, the system offers a new level of home security for entry doors, swimming pools, and so on. These sentry cameras can be equipped to transmit an image of the person being filmed directly to a TV screen inside the house, or to record an instant photo marked with the time it was taken, thus providing a record of all callers.

Looking a bit further into the future, we see homes that have even more electronic capabilities for security, comfort, and convenience. Picture the modern homeowner awakening to music from unseen speakers. A timer automatically opens the curtains to let the morning sun shine in, and the room has warmed up to a preselected temperature after a night of energy-saving with the thermostat turned back. A touch of a button starts the shower at the temperature this homeowner prefers, while in the kitchen the coffee pot turns on and as he steps into the room the TV clicks on to a favorite morning station.

Throughout the day the central control system automatically monitors and protects this home, heating and cooling in an energy-efficient manner, watering the lawn at the most opportune time, even starting the whirlpool as the homeowner drives in from work.

A model community recently constructed near Orlando, Florida, features an entire interactive community. Fiber optic cable links individual homes with the hospital, police department, and cable TV system. This feature allows communications, security, and entertainment needs to be met in a unique network. For example, "Medicalert" sensors installed throughout the homes permit residents to contact the hospital immediately if they need help.

Eventually residents will be able to order groceries from a local market without leaving home, choosing individual products from a series of video images flashed onto the TV screen. Home shoppers will be able to switch from the produce aisle to the dairy case at the touch of a button. When the shopping is done, payment is made via an electronic funds transfer from the buyer's bank account to that of the supermarket.

The heart of all such monitoring systems is a central processor tied to a computer monitor. Functions are programmed, canceled, or changed at the touch of a fingertip, using a floor plan of the home on a touch-sensitive screen, plus step-by-step instructions. As the technology evolves, prices for such systems will come down. In the early 1990s, more than a dozen electronics firms were marketing some version of a computer-controlled home control system. Installing such equipment will be all in a day's work for the residential electricians of tomorrow.

The scientific field known as robotics is likely to have an impact on construction's future as well as home and office design. Robotic arms are already employed in many industries, lifting and manuevering heavy loads according to a predetermined program, and performing tasks considered too repetitive or too dangerous for human workers. Robotic vacuum cleaners and lawnmowers will eventually come to the consumer market, leading the way for a legion of electronic servants. Who is to say science will not produce a robotic power nailer or painting arm?

An increasing emphasis on structures that look well yet work hard and require little maintenance will continue to bring new products to the marketplace. As a forward-thinking person, you will want to learn about

Toward the Future

these products and learn how to use them on the job. In this way you will remain a flexible worker, ready to adapt to change.

By now you know there is much more to the construction industry than mere hammers and nails. An exciting future awaits those willing to develop the attitude, skills, and knowledge that can open doors to a rewarding and challenging world of work. If you want that world for your own, you now know how to take the first step.

Glossary

apprentice Person, usually between 18 and 24 years of age, who is learning a trade through a combination of on-the-job training and classroom instruction.

agreement, collective bargaining Contract negotiated between a union and an employer to cover workers' wages, hours, fringe benefits, and working conditions.

building system Used interchangeably with the term "systems construction," in which materials, prefabricated units, and labor are scheduled for use to maximize efficiency.

blueprint Drawing reproduced by a photographic process depicting an architect's or designer's construction drawing.

building trades Skilled trades represented in the construction industry, including bricklayers, carpenters, electricians, masons, painters, and plumbers.

contractor Individual or company who agrees to do a specific job under conditions and prices spelled out in a legal agreement called a contract.

craftsperson Artisan whose work or occupation requires particular training and practice to achieve a high level of expertise.

estimator Worker who calculates the amount of materials, labor, and general costs required to accomplish a job.

fabrication The act of constructing an item from standardized parts; components that arrive at a con-

GLOSSARY

struction site ready for installation are said to be fabricated.

finish work The final, sometimes decorative, work involved as construction reaches its final stages; work usually requiring expert care and skill.

foreperson Leader of a work crew, usually specially trained or more experienced.

form Framework of wood or metal into which concrete is poured to achieve a desired shape.

frame Parts fitted together to achieve a desired shape, usually of wood in construction projects.

fringe benefits Benefits over and above the basic wage rate; for example, health insurance coverage, paid vacations, etc.

inspector Person who examines a project or work element to guarantee that standards of safety or quality are being met.

joiner Specialist in building structural elements of joined wood, particularly doors and windows.

joists Horizontal beams used to support a floor, roof, or ceiling.

journeyperson Worker who has completed apprenticeship and is prepared for full responsibility and earning potential in a construction job; formerly called journeyman.

lamination Process by which layers of wood or other materials are held together with resin or adhesives and bonded under pressure or heat.

layoff Interruption in employment usually caused by a construction slowdown or temporary halt at a worksite because of inclement weather.

layout work Reading of blueprints and specifications and translating the information into instructions for workers.

manual skills Those performed by hand, requiring the use of skill or energy.

Glossary

millwork Finished carpentry work brought to a site for installation; for example, window frames or prehung doors.

on-the-job training Paid employment that combines work with learning.

overtime Time spent on the job beyond the basic workday or workweek as defined by law. Premium rates of pay apply to such hours.

power tool Any tool requiring other than human power to operate, generally electric.

prefabrication Manufacture of construction elements in a factory, speeding construction assembly at the job site.

rafters Parallel beams that support a roof.

reinforced concrete Concrete into which steel mesh or rods are embedded to increase strength.

resilient In building materials, describing flooring products of asphalt or vinyl composition, designed to absorb shock and sound.

rough framing The skeletal framework of a building, usually constructed of dimensional lumber or steel.

scaffold Temporary or movable framework or platform from which workers can reach work sites above ground level.

shop course School study that emphasizes the use of tools in a trade or craft.

standard General model of safety, quality, or performance to which other work is compared.

subcontractor Person or company that agrees to perform certain skilled work on a building being erected by a general contractor.

superintendent Supervisor who directs the work of forepersons and their crews at a construction site.

troubleshooting Term applied to the process of testing and/or diagnosing malfunctions in electrical circuits, appliances, machinery, or other equipment.

GLOSSARY

trim The final stages of a woodworking or construction project, requiring the highest level of performance; also the final stages of equipping a building with woodwork, cabinetry, doors, and so on.

wage scale Listing of wages from highest to lowest, comparing compensation for specific job categories.

Appendix

Colleges Offering Construction Degree Programs

Arizona State University
Division of Construction
Tempe, AZ 85281

Auburn University
Department of Building Technology
Auburn, AL 36830

Bradley University
Department of Construction
Peoria, IL 61606

California Polytechnic State University
School of Architecture and Environmental Design
San Luis Obispo, CA 93401

California State University—Fresno
Department of Industrial Arts and Technology
Fresno, CA 93726

Clemson University
Department of Building Science
Clemson, SC 29631

University of Colorado
College of Engineering and Applied Science
Boulder, CO 80302

APPENDIX

Drexel University
Construction Management
Philadelphia, PA 19104

University of Florida
Department of Building Construction
Gainesville, FL 32601

Georgia Institute of Technology
School of Architecture
Atlanta, GA 30332

Hampton Institute
Building Construction Engineering
Hampton, VA 23368

University of Houston
Department of Civil Technology
Houston, TX 77004

Indiana University—Purdue
Department of Construction Technology
Indianapolis, IN 46202

Iowa State University
Construction Engineering
Ames, IA 50010

John Brown University
Department of Building Construction
Siloam Springs, AK 72761

Kansas State University
Department of Construction Science
Manhattan, KS 66502

APPENDIX

Kansas State College of Pittsburg
Building Technology Department
Pittsburg, KS 66762

Louisiana State University
Construction Technology Curriculum
Baton Rouge, LA 70803

Memphis State University
Department of Engineering Technology/Construction
Memphis, TN 38111

Southwest Missouri State College
Department of Industrial Education
Springfield, MO 65802

University of Nebraska
Department of Construction Management
Lincoln, NE 68508

Northeast Louisiana University
Department of Building Construction
Monroe, LA 71201

Oklahoma State University
Construction Management Technology
Stillwater, OK 74074

Purdue University
Department of Construction Technology
West Lafayette, IN 47967

Texas A&M University
Department of Building Construction
College Station, TX 77843

Virginia Polytechnic Institute
Department of Building Construction
Blacksburg, VA 24061

University of Washington
Department of Building Construction
Seattle, WA 98105

Washington State University
Building Theory and Practice
Pullman, WA 99163

West Virginia State College
Department of Light Construction
Institute, WV 25112

University of Wisconsin—Platteville
College of Industry
Platteville, WI 53818

University of Wisconsin—Stout
School of Applied Sciences and Technology
Menomonie, WI 54751

COLLEGE AND UNIVERSITY CHAPTERS, NATIONAL ASSOCIATION OF HOME BUILDERS
Established in 1971, the NAHB Student Chapter program offers young men and women enrolled in construction and related courses the opportunity to receive special membership benefits. By interacting with local builder associations, members get a first-hand look at the industry they are preparing to enter. They also work side by side with professional builders on community projects and benefit from NAHB-sponsored scholar-

ships. This list is confined to colleges and universities, but there are also NAHB student chapters in many American high schools and vocational-technical schools.

ALABAMA
Jefferson State University
Department of Technology and Engineering
Birmingham, AL 35215-3098

ARIZONA
Arizona State University
College of Architecture
Tempe, AZ 85287

CALIFORNIA
California State University—Chico
Construction Management
Chico, CA 95929-0305

Orange Coast College
Construction Technology
Costa Mesa, CA 92626

California State University—Fresno
Department of Industrial Arts
Fresno, CA 93740

Fresno City College
Fresno, CA 93741

California Institute of Technology—Fullerton
Fullerton, CA 92631

San Diego State University
San Diego, CA 93182

California State Polytechnic University—Pomona
Engineering Technical Department
Pomona, CA 91768

COLORADO
Colorado State University
Department of Industrial Sciences
Ft. Collins, CO 80523

University of Denver
Department of Construction Management
Denver, CO 80208

FLORIDA
University of Florida
School of Building Construction
Gainesville, FL 32611

University of North Florida
Division of Technologies
Jacksonville, FL 32216-6699

GEORGIA
Georgia Southern University
School of Technology
Statesboro, GA 30460-8044

Gwinnett Technical Institute
Lawrenceville, GA 30245

IDAHO
Ricks College
Construction Management Department
Rexburg, ID 83460-1030

APPENDIX

ILLNOIS
Illinois State University
Department of Industrial Technology
Normal, IL 61761

Triton College
River Grove, IL 60171

Rock Valley College
Rockford, IL 61111

Southern Illinois University—Carbondale
Applied Technology-Construction
Carbondale, IL 62901

Belleville Area College
Construction Management
Belleville, IL 62221

INDIANA
Indiana State University
School of Technology
Terre Haute, IN 47802

Vincennes University
Technology Department
Vincennes, IN 47591

Purdue University
Department of Building Construction
West Lafayette, IN 47907

IOWA
Des Moines Area Community College
Ankeny, IA 50021

KENTUCKY
Northern Kentucky University
Technology Department
Highland Heights, KY 41076

LOUISIANA
Louisiana State University
Department of Industrial and Technical Education
Baton Rouge, LA 70803

MAINE
Eastern Maine Technical College
Bangor, ME 04401

MARYLAND
Montgomery College
Rockville, MD 20850

MASSACHUSETTS
Wentworth Institute of Technology
Boston, MA 02115

MICHIGAN
Delta College
University Center, MI 48710

Central Michigan University
Department of Industrial and Engineering Technology
Mt. Pleasant, MI 48859

Ferris State University
Associated Construction
Big Rapids, MI 49307

Michigan State University
East Lansing, MI 48824-1323

Appendix

Western Michigan University
Construction Science and Management
Kalamazoo, MI 49008

MINNESOTA
Moorhead State University
Department of Industrial Studies
Moorhead, MN 56560

MISSISSIPPI
Hinds Junior College
Raymond, MS 39154

University of Southern Mississippi
Hattiesburg, MS 39406

MISSOURI
University of Missouri
Columbia, MO 65201

MONTANA
Montana State University
Department of Agriculture and Industrial Education
Bozeman, MT 59717-0374

NEBRASKA
Northeast Community College
Norfolk, NE 68701

Kearney State College
Kearney, NE 68847

Southeast Community College
Milford, NE 68405

APPENDIX

NEVADA
University of Nevada—Las Vegas
Architecture and Applied Studies
Las Vegas, NV 89154

NEW HAMPSHIRE
Cheshire Vocational Center
Keene, NH 03431

Huot Vocational Education Center
Laconia, NH 03246

NEW YORK
Utica College of Syracuse University
Construction Management Department
Utica, NY 13502

Hudson Valley Community College
Department of Civil Engineering/Construction Technology
Troy, NY 12180

NORTH CAROLINA
Appalachian State University
Department of Technology
Boone, NC 28608

Coastal Carolina Community College
Jacksonville, NC 28546

NORTH DAKOTA
North Dakota State University
Fargo, ND 58105

APPENDIX

OHIO
   Bowling Green State University
   College of Technology
   Bowling Green, OH 43403

   Ohio State University
   Agricultural/Technical Institute
   Wooster, OH 44691

   Apollo Career Center
   Lima, OH 45806

OKLAHOMA
   University of Oklahoma
   College of Architecture
   Norman, OK 73019

   Oklahoma State University
   Department of Construction Management
   Stillwater, OK 74078

OREGON
   Lane Community College
   Eugene, OR 97405

   Oregon State University
   Corvallis, OR 97331-5101

   Sabin Occupational Skills Center
   Milwaukee, OR 97267

PENNSYLVANIA
   Pennsylvania State University
   Department of Engineering
   University Park, PA 16802

APPENDIX

SOUTH CAROLINA
Clemson University
Department of Building Science
College of Architecture
Clemson, SC 29634-0507

SOUTH DAKOTA
Western Dakota Vocational Technical Institute
Rapid City, SD 57701-4178

Lake Area Vocational Technical Institute
Watertown, SD 57201

TEXAS
North Lake College
Technology Division
Irving, TX 75038-3899

Richland College
Dallas, TX 75243

San Antonio College
San Antonio, TX 78211

Texas A&M University
Department of Construction Science
College Station, TX 77843-3137

Texas State Technical Institute
Building Costruction Technology
Waco, TX 76705

University of Texas at San Antonio
College of Fine Arts and Humanities Architecture
   Program
San Antonio, TX 78285-0642

Appendix

UTAH
Brigham Young University
Provo, UT 84602

Bridgerland Applied Technology Center
Logan, UT 84321

Snow College
Ephraim, UT 84627

Utah Valley Community College
Orem, UT 84058

Weber State College
Ogden, UT 84408

VIRGINIA
Virginia Polytechnic Institute
Constructors Consortium
Blacksburg, VA 24061

WASHINGTON
University of Washington
Department of Building Construction
Seattle, WA 98195

WEST VIRGINIA
Ben Franklin Career Center
Dunbar, WV 25064

WISCONSIN
North Central Technical College
Wausau, WI 54401-1899

Wisconsin Indianhead Technical Institute
Rice Lake, WI 54868

University of Wisconsin—Madison
Madison, WI 53706

University of Wisconsin—Platteville
Industrial Studies
Platteville, WI 53818-3099

University of Wisconsin—Stout
Menomonie, WI 54751

For Further Reading

The following reference books, periodicals, brochures, and pamphlets are available at school and public libraries, through a school guidance counselor, or by mail. They are published by organizations interested in promoting careers in the construction trades.

American Electricians' Handbook: Reference Book for the Practical Electrical Man. 9th ed. Terrell Croft. McGraw-Hill, New York, 1970.

Apprenticeship: Past and Present, free. Bureau of Apprenticeship and Training Administration, U.S. Department of Labor, 200 Constitution Avenue NW, Washington, DC 20210.

Basic Plumbing Illustrated. Sunset Books, Menlo Park, CA, 1975.

Big Questions, and *Reach Out for Tomorrow*, free. National Joint Painting, Decorating and Drywall Apprenticeship and Training Committee, 1750 New York Avenue NW, Washington, DC 20006.

Career Choice and Job Search (a workbook). Jay Como. Meridian Educational Publishers, 1986.

Career Choices: A Guide for Teens and Young Adults: Who Am I? What Do I Want? How Do I Get It? Mindy Bingham and Sandy Stryker. Able Publishing, 1990.

Careering: Charting Your Own Future. Neil M. Yeager, John Wiley & Sons, Inc., New York, 1988.

Career Planning: Freedom to Choose, 3d ed. Bruce E. Shertzer. Houghton-Mifflin, 1990.

Careers in Masonry, free. National Concrete Masonry Association, P.O. Box 781, Herndon, VA 22070.

Careers in Sheet Metal, free. National Fund, Sheet Metal and Air Conditioning Industry, attention: Careers, 601 North Fairfax Street, Alexandria, VA 22314.

Careers in the Crafts, single copy free. International Association of Machinists and Aerospace Workers, Research Library, 1300 Connecticut Avenue NW, Washington, DC 20036.

Carpenter, Yellow Pages Career Library, booklet sponsored by the National Association of Elementary School Principals. Murphy Levy Wurman, 1214 Arch Street, Philadelphia, PA 19107.

Carpentry as an Occupation, free. United Brotherhood of Carpenters and Joiners of America, 101 Constitution Avenue NW, Washington, DC 20001.

Carpentry: A Trade Worth Learning Through Apprenticeship, single copy free. U.S. Department of Labor, Employment and Training Administration, Washington, DC 20213.

College Plus: Put Your Degree to Work with Trade and Technical Skills, free. National Association of Trade and Technical Schools, 2251 Wisconsin Avenue NW, Washington, DC 20016.

Construction: A Guide for the Profession. Thomas A. Grow. Prentice-Hall, Englewood Cliffs, NJ, 1975.

Construction Contracting, 3d ed. Richard H. Clough. Wiley-Interscience, New York, 1975.

Construction Foreman's Job Guide. James E. Clyde, John Wiley & Sons, Inc., New York, 1987.

Construction: Industry and Careers. William P. Spence. Prentice-Hall, New York, 1990.

Dwelling House Construction, 4th ed. Albert G.H. Dietz. MIT Press, Cambridge, MA, 1974.

For Further Reading

Exploring Nontraditional Jobs For Women. Rose Neufeld. The Rosen Publishing Group, Inc., New York, 1989.

High-Paying Blue-Collar Jobs for Women: A Comprehensive Guide. Larry J. Ricci. Ballantine Books, New York, 1981.

How to Become the Successful Construction Contractor. Taylor F. Winslow. Craftsman Book Co., Solana Beach, CA, 1975.

Masons and Builders Library. Louis M. Dezettel. Audel, Indianapolis, IN, 1972.

Occupational Outlook Handbook (sections on all major job categories including detailed job descriptions, hiring projections, and earnings potential). Bureau of Labor Statistics, U.S. Department of Labor.

Opportunities in Building Construction Trades. Michael Sumichrast. National Textbook Company, Lincolnwood, IL 60646, 1985.

Opportunities in Carpentry. Roger Sheldon. National Textbook Co., Skokie, IL, 1980.

Opportunities in Crafts Careers. Marianne F. Munday. National Textbook Co., Lincolnwood, IL 60646, 1988.

Opportunities in Landscape Architecture. Thomas A. Griswold and William G. Swain. National Textbook Co., Skokie, Ill, 1978.

Index

A
advancement, opportunities for, 29, 32, 33, 35, 40, 43, 53, 89, 93–98
agreement, apprenticeship, 75
American Federation of Labor–Congress of Industrial Organizations, 61
American Society of Home Inspectors, 27
American Society of Landscape Architects, 43
apprentice, 7, 13, 18, 25, 37, 45
apprenticeship, 2, 7, 18, 19, 23, 25, 27, 28, 30, 33, 39, 43, 45–46, 49–50, 52, 63–64, 68, 72–80, 104
  industry, 70
  nonunion, 52, 123
  union, 33, 52, 56, 58, 62, 63
aptitude testing, 7, 72–73, 91–92
architect, 14, 33
architectural woodworking, 25
asbestos worker, 10
Associated General Contractors of America, 79

B
bidding, competitive, 106–107
bricklayer, 7, 9, 11, 17–19, 57–58, 72
building construction technology, 81–82
building inspector, 96–97
Building Trades Union, 66
Burt Hill Kosar Rittelman Associates, 146
business, owning your own, 7, 24, 29, 33, 39, 43, 50, 70, 80, 106–115

C
cabinetmaker, 24–25
carpenter, 7, 9, 11, 12–14, 16, 19–25, 80, 82
  apprenticeship, 64, 75, 77–79, 140–142
cement mason, 16, 18–19
COLA (cost-of-living adjustment), 61
color vision, 31, 123
commercial construction, 14, 35, 41, 49, 57, 61, 80, 82, 106
CommuniCraft, 5
community college, 7, 42, 68, 80–84
community, interactive, 151–152
Community Revitalization Project, 70
computer, 26, 40–41
concrete mason, 9
condominiums, 2, 3, 21
construction carpentry, 82
construction inspector, 25–27
construction technology, 81

175

INDEX

contractor, 8, 14, 21, 26, 29, 41, 44, 75, 81
  electrical, 31
  general, 17, 47, 54
  landscape, 42
  self-employed, 7, 107–108
  specialty, 35
controls, electronic, 31, 32, 146
core systems, 147
Correctional Training Program, 70
Craft Math, 5
Craft Skills Program, 66, 70
crew chief, 93–94
Critical Path Method, 144
custom builder, 109, 119–120

D

degree, 68
  associate, 80–84
  bachelor's, 81, 84–85
  four-year, 43–44, 69
  master's, 84–85
  two-year, 27, 69
Dickens, Michael, 147, 149
downtime, 23, 29, 35, 46, 53, 87, 99, 100
drywall mechanic, 10, 28–30, 124–126
dues, union, 60–61

E

education, construction, 68–85
electrical inspector, 97
electrician, 7, 9, 11, 30–32, 75, 80, 82–83, 107, 121–124
  apprenticeship, 64, 75–77
elevator constructor, 9–10
elevator inspector, 97
engineer, 10
  civil, 95–96
entrepreneurship, 106–115
estimating, 33, 81, 84, 107, 112
estimator, 33, 94–95
expediter, 95

F

factory-built structure, 148–149
fieldwork, 26, 35–36, 40–41
finish
  carpentry, 19–21, 120–121
  electrical work, 30–32, 121–123
finishing construction, 10
floor-covering installer, 10, 29, 34
floor layer, 33
fringe benefits, 32, 39, 40, 50, 51, 56, 89

G

G.E. Plastics, 146, 149
glazier, 16, 35–36
glue-lamination, 150

H

hazards, job, 20, 27, 30–31, 35, 38, 45, 48, 52, 55, 57, 59–60, 100
heating air-conditioning, refrigeration mechanic, 10–11, 16, 36–37, 84
helper, 9, 11, 18, 28, 39, 42
high school, 4, 5, 7, 45, 68, 103–104
Home Builders Institute, 5, 66, 71, 72
home inspector, 97–98

I

IBACOS, Inc., 145–148
illiteracy, workplace, 4
industrial construction, 30, 39
inspector, 25–27, 96–98
insulation worker, 16, 39
interior systems carpenter, 19, 24
International Association of Bridge, Structural and Ornamental Iron Workers, 56
International Association of Heat and Frost Insulators, 39
International Association of Plumbing and Mechanical Inspectors, 27

INDEX

International Brotherhood of Bricklayers and Allied Craftsmen International, 58
International Brotherhood of Electrical Workers (IBEW), 32, 59–60, 77
International Brotherhood of Painters and Allied Trades (IBPAT), 30, 51, 137
ironworker, 9, 54–56

J
Job Corps, 7, 42, 60, 64–65, 68, 70–72, 139
jobs applying for, 88–92
Joint Apprentice Committees, 46, 79
journeyworker, 9, 13, 25, 46, 56, 62, 72, 75, 88, 96, 106, 122

L
laborer, 9, 39–40
Laborers' International Union of North America, 40
Labor, U.S. Department of, 31, 45, 73, 74, 79
landscape architect, 42–44
landscaper, 16, 72, 80, 126–129
land surveyor, 40–42
large electrical work, 30
lather, 9, 28–30
licensing, 27, 32, 42, 52, 95–96

M
machine operator, 39, 44–48, 54–55
maintenance, 3, 14, 23, 48–49, 50, 52
marble installer, 9, 57–58
marketing, 50, 111–112
Masco Corp., 146
materials handling, 25, 39–40
mechanical construction, 9

mechanical inspector, 97
millwright, 24–25
minorities, in construction, 63–67, 73

N
National Association of Home Builders (NAHB), 4, 10, 70, 79, 85
National Association of Women in Construction (NAWIC), 67, 115
National Electrical Code, 30, 83, 97, 122
National Urban League, 67

O
on-the-job training (OJT), 7, 19, 33, 39, 42, 46, 54, 68–69
operating engineer, 7, 9, 44–48, 100–101
Oriental Strand Board, 149
ornamental ironworking, 54
overtime, 4, 11, 30, 52, 100

P
painter, 10, 16, 48–51, 72, 129–138
paperhanger, 48–51
payscale, 2, 3, 16–17, 19, 23, 25, 27, 30, 32, 34, 36, 39, 40, 43–44, 45, 51, 54, 58, 96, 98
  apprentice, 6, 30, 73
  nonunion, 42, 50, 56, 126
pipefitter, 10, 11, 52
plasterer, 10, 138–139
plumber, 7, 9, 10, 27, 51–53, 72, 80, 83–84
plumbing inspector, 97
project manager, 94
public works inspector, 97

Q
qualifications, job, 102–103

177

INDEX

R

recruitment, women and minorities, 66–67, 123
reinforcing ironworker, 56
remodeling, 5, 9, 20, 23, 33, 82, 108, 109
research, homebuilding, 145–148
residential construction, 14, 21, 29, 30, 39, 81, 106–107
rigger, 9, 54
robotics, 152–153
roofer, 9, 10, 53–54, 107
rough carpentry, 20, 120
rough electrical work, 30–31, 120
Ryland Homes, 146

S

safety equipment, 12, 30–31, 120
safety, workplace, 59–60, 73, 123
scarfito plaster carving, 138–139
self-employment, 8, 17, 23, 24, 32, 41, 42, 43, 50, 52, 83, 106–115, 119
Service Corps of Retired Executives (SCORE), 115
sheetmetal worker, 9
space station, 9, 150
speculative builder, 109–110
steamfitter, 10
stonemaston, 9, 10, 17–19, 57–58
structural construction, 9
structural steel workers, 54–56
sub-pay, 61, 87, 101
superintendent
  general, 94
  job, 13, 81, 94, 107
supervisor, 13, 19, 20, 24, 26, 29, 32, 33, 50, 53, 80, 81, 93–98
survey technician, 40–42, 95
system integration, 147

T

technical institute, 68, 81–84
terrazzo installer, 9, 57–58
tile installer, 8, 57–58
tools, 12, 72, 121
  hand, 17, 20, 24, 29, 31, 35, 38, 52, 57, 82
  power, 20–21, 23, 35, 51, 82, 121
trade school, 7, 68, 69, 70–71
training, 5–6, 32, 49, 68–85
  preapprenticeship, 39, 68, 69

U

union membership, 14, 21, 34, 56, 58, 59–62, 88, 101, 123
Union of Operating Engineers, 48
United Brotherhood of Carpenters and Joiners of America, 30, 58, 79, 140
USG Corp., 146

V

vocational training, 7, 65, 68–69, 103, 134

W

wall-covering installers, 50, 135–136
weather, 41, 52, 54, 61, 88
  delays, 14, 17, 35, 48, 100
Wider Opportunities for Women (WOW), 66–67
women in construction, 2, 63–67, 75, 103
working conditions, 17, 20, 22, 29, 30–31, 33, 35, 38, 39, 41, 45, 48, 52, 53, 55, 88, 102, 123

Y

Young Women's Christian Association (YWCA), 67